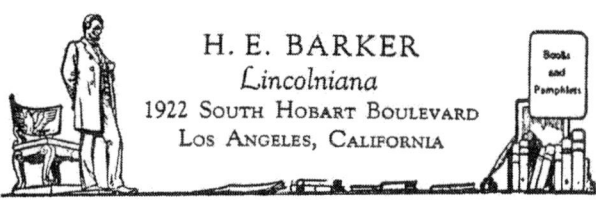

DAY'S ALGEBRA.
****

Three works on Mathematics are known to have interested Abraham Lincoln. During his brief school days he studied Pike's Arithmetic.

While riding the Circuit in the 40s he carried with him a copy of Euclid.

Just when he studied Day's Algebra is uncertain, but his own copy with his name on its fly-leaf, is owned by the Chicago Historical Society, and as it is an edition of 1647, it is more than likely that he studied both Geometry and Algebra during the same period.

H. E. Barker

AN

INTRODUCTION

TO

A L G E B R A,

BEING

THE FIRST PART

OF

A COURSE OF MATHEMATICS,

ADAPTED TO THE METHOD OF INSTRUCTION IN
THE AMERICAN COLLEGES.

———————

BY JEREMIAH DAY,

Professor of Mathematics and Natural Philosophy, in
Yale College.

———————

NEW-HAVEN,
PUBLISHED BY HOWE & DEFOREST.
....
OLIVER STEELE, PRINTER.
....
1814.

mg 5233

PREFACE.

THE following summary view of the first principles of algebra, is intended to be accommodated to the method of instruction generally adopted in the American colleges.

The books which have been published, in Great-Britain, on mathematical subjects, are principally of two classes.—One consists of extended treatises, which enter into a thorough investigation of the particular departments which are the objects of their inquiry. Many of these are excellent in their kind; but they are too voluminous for the use of the body of students in a college.

The other class are expressly intended for beginners; but many of them are written in so concise a manner, that important proofs and illustrations are excluded. They are mere *text-books*, containing only the *outlines* of subjects which are to be explained and enlarged upon, by the professor in his lecture room, or by the private tutor in his chamber.

In the colleges in this country, there is generally put into the hands of a class, a book from which they are expected *of themselves* to acquire the principles of the science to which they are attending; receiving, however, from their instructor, any additional assistance which may be found necessary. An elementary work for such a purpose, ought evidently to contain the explanations which are requisite, to bring the subjects treated of within the comprehension of the body of the class.

If the design of studying the mathematics were merely to obtain such a knowledge of the *practical* parts, as is required for transacting business; it might be sufficient to commit to memory some of the principal rules, and to make the operations familiar, by attending to the examples. In this mechanical way, the accountant, the navigator, and the land-surveyor, may be qualified for their respective employments, with very little knowledge of the *principles* that lie at the foundation of the calculations which they are to make.

But a higher object is proposed, in the case of those who are acquiring a liberal education. The main design should be to call into exercise, to discipline, and to invigorate the powers of the mind. It is the *logic* of the mathematics which constitutes their principal value, as a part of a course of collegiate instruction. The time and attention devoted to them, is for the purpose of forming *sound reasoners*, rather than expert mathematicians. To accomplish this object, it is necessary that the principles be clearly explained and demonstrated, and that the several parts be arranged in such a manner, as to show the dependence of one upon another. The whole should be so conducted, as to keep the reasoning powers in continual exercise, without greatly fatiguing them. No other subject affords a better opportunity for exemplifying the rules of correct thinking. A more finished specimen of clear and exact logic has, perhaps, never been produced, than the Elements of Geometry by Euclid.

It may be thought, by some, to be unwise to form our general habits of arguing, on the model of a science in which the inquiries are accompanied with *absolute certainty;* while the common business of life must be conducted upon *probable* evidence, and not upon principles which admit of complete demonstration. There would be weight in this objection, if the attention were confined to the *pure* mathematics. But when these are connected with the *physical* sciences, astronomy, chemistry, and natural philosophy, the mind has opportunity to exercise its judgment, upon all the various degrees of probability which occur in the concerns of life.

So far as it is desirable to form a *taste* for mathematical studies, it is important that the books by which the student is first introduced to an acquaintance with these subjects, should not be rendered obscure and forbidding by their conciseness. Here is no opportunity to awaken interest, by rhetorical elegance, by exciting the passions, or by presenting images to the imagination. The beauty of the mathematics depends on the distinctness of the objects of inquiry, the symmetry of their relations, the luminous nature of the arguments, and the certainty of the conclusions. But how is this beauty to be perceived, in a work which is so much abridged, that the chain of reasoning is often interrupted, many difficulties left unexplained, important demonstrations omitted, and the transitions from one subject to another so abrupt, as to keep their connections and dependencies out of view?

It may not be necessary to state every proposition and its proof, with all the formality which is so strictly adhered to by Euclid; as it is not essential to a logical argument, that it be expressed in regular and entire syllogisms. A step of a demonstration may be safely left out, when it is so simple and obvious, that no one possessing a moderate acquaintance with the subject, could fail to supply it for himself. But this liberty of omission ought not to be extended, to cases in which it might occasion obscurity and embarrassment. If it be desirable to give opportunity for the mind to display and enlarge its powers, by surmounting obstacles; full scope may be found for this kind of exercise, especially in the higher branches of the mathematics, from difficulties which will unavoidably occur, without creating new ones for the sake of perplexing.

The purpose for which abridged compilations are commonly made is, probably, to *save time*. The *expense* of an additional volume or two, in that part of a public education which is to occupy a large portion of three or four years, can hardly be supposed to be an object of great comparative importance. The principal saving of expense, in this case, is included in the saving of time. But is not the progress of the student impeded, rather than accelerated, by abridgments? The time requisite to become master of a subject, is not always proportioned to the number of *pages* which it occupies. Hours may be spent, in supplying an explanation, or an article of proof, which, if it had been inserted in its place, might have been read and understood, in a few minutes.

Algebra requires to be treated in a more plain and diffuse manner, than some other parts of the mathematics; because it is to be attended to, *early* in the course, while the mind of the learner has not been habituated to a mode of thinking so abstract, as that which will now become necessary. He has also a *new language* to learn, at the same time he is settling the *principles* upon which his future inquiries are to be conducted. These principles ought to be established, in the most clear and satisfactory manner which the nature of the case will admit of. Algebra and geometry may be considered as lying at the foundation of the succeeding branches of the mathematics, both pure and mixed. Euclid and others have given to the geometrical part, a degree of clearness and precision which would be very desirable, but is hardly to be expected, in algebra.

For the reasons which have been mentioned, the manner in which the following pages are written, is not the most concise. But the work is necessarily limited in extent of subject. It is far from being a *complete* treatise of alge- bra. It is merely an introduction. It is intended to con- tain as much matter, as the student at college can attend to, with advantage, during the short time allotted to this partic- ular study. There is generally but a small portion of a class, who have either leisure or inclination, to pursue mathematic- al inquiries much farther, than is necessary to maintain an honourable standing, in the institution of which they are mem- bers. Those few who have an unusual taste for this science, and aim to become adepts in it, ought to be referred to sepa- rate and complete treatises, on the different branches. No one who wishes to be thoroughly versed in mathematics, should look to compendiums and elementary books, for any thing more, than the first principles. As soon as these are acquired, he should be guided in his inquiries, by the genius and spirit of original authors.

In the selection of materials, those articles have been ta- ken which have a practical application, which are not of very difficult comprehension, and which are preparatory to succeeding parts of the mathematics, philosophy, and astron- omy. The object has not been, to introduce *original matter.* In the mathematics, which have been cultivated with success, from the days of Pythagoras, and in which the principles al- ready established are sufficient to occupy the most active mind for years, the parts to which the student ought *first* to attend, are not those recently discovered. Free use has been made of the works of Newton, Maclaurin, Saunderson, Simpson, Euler, Emerson, Lacroix, and others, but in a way that rendered it inconvenient to refer to them, in particular instances. The proper field for the display of mathematical *genius,* is in the region of invention. But what is requisite for an elementary work, is to collect, arrange, and illustrate, materials already provided. However humble this employ- ment, he ought patiently to submit to it, whose object is to instruct, not those who have made considerable progress in the mathematics, but those who are just commencing the study. Original discoveries are not for the benefit of *begin- ners,* though they may be of great importance to the ad- vancement of science.

The arrangement of the parts is such, that the explanation of one is not made to depend on another which is to follow. The addition, multiplication, and division of *powers,* for in- stance, is placed after involution. In the statement of gene-

ral rules, if they are reduced to a small number, their applications to particular cases may not, always, be readily understood. On the other hand, if they are very numerous, they
become tedious and burdensome to the memory. The rules
given, in this introduction, are most of them comprehensive;
but they are explained and applied, in subordinate articles.

A *particular* demonstration is sometimes substituted for a
general one, when the application of the principle to other
cases is obvious. The examples aro not often taken from
philosophical subjects, as the learner is supposed to be familiar with none of the sciences except arithmetic. In treat-
ing of *negative* quantities, frequent references are made to
mercantile concerns, to debt and credit, &c. These are merely for the purpose of illustration. The whole doctrine of
negatives is made to depend on the single principle, that
they are quantities to be *subtracted*. But the student, at
this early period, is not accustomed to abstraction. He requires particular examples, to catch his attention, and aid his
conceptions.

The section on *proportion* will, perhaps, be thought useless to those who read the fifth book of Euclid. That is
sufficient for the purposes of pure *geometrical* demonstration.
But it is important that the propositions should also be presented, under the algebraic forms. In addition to this, great
assistance may be derived from the algebraic *notation*, in demonstrating, and reducing to system, the laws of proportion.
The subject, instead of being broken up into a multitude of
distinct propositions, may be comprehended in a few general principles.

CONTENTS.

INTRODUCTORY OBSERVATIONS

MATHEMATICS in GENERAL.

Art. 1. MATHEMATICS *is the science of* QUANTITY.

Any thing which can be *increased* or *diminished*, or which is capable of being *measured*, is called *quantity*. Thus, a *line* is a quantity, because it can be made longer or shorter; and can be measured, by applying to it another line, as a foot, a yard, or an ell. *Weight* is a quantity, which can be measured, in pounds, ounces, and grains. *Time* is a species of quantity, whose measure can be expressed, in hours, minutes, and seconds. But *colour* is not a quantity. It cannot be said, with propriety, that one colour is either greater or less than another. The operations of the *mind*, such as thought, choice, desire, hatred, &c. are not quantities. They are incapable of mensuration.

2. Those parts of the Mathematics, on which all the others are founded, are *Arithmetic*, *Algebra*, and *Geometry*.

3. Arithmetic is the science of *numbers*. Its aid is required, to complete and apply the calculations, in almost every other department of the mathematics.

B

4. ALGEBRA is a method of computing principally by *letters*. FLUXIONS may be considered as belonging to the higher branches of algebra.

5. GEOMETRY is that part of the mathematics, which treats of *magnitude*. By magnitude, in the appropriate sense of the term, is meant that species of quantity, which is *extended*; that is, which has one or more of the three dimensions, *length, breadth*, and *thickness*. Thus, a *line* is a magnitude, because it is extended, in length. A *surface* is a magnitude, having length and breadth. A *solid* is a magnitude, having length, breadth, and thickness. But *motion*, though a quantity, is not, strictly speaking, a magnitude. It has neither length, breadth, nor thickness.*

6. TRIGONOMETRY and CONIC SECTIONS are branches of the mathematics, in which, the principles of geometry are applied to *triangles*, and the sections of a *cone*.

7. Mathematics are either pure, or mixed. In *pure* mathematics, quantities are considered, independently of any substances actually existing. But, in *mixed* mathematics, the relations of quantities are investigated, in connection with some of the properties of matter, or with reference to the common transactions of business. Thus, in Surveying, mathematical principles are applied to the measuring of land; in Optics, to the properties of light; and in Astronomy, to the motions of the heavenly bodies.

8. The science of the pure mathematics has long been distinguished, for the clearness and distinctness of its principles; and the irresistible conviction, which they carry to the mind of every one who is once made acquainted with them. This is to be ascribed, partly to the nature of the subjects, and partly to the exactness of the definitions, the axioms, and the demonstrations.

* NOTE. Some writers, however, use magnitude, as synonymous with quantity.

9. The foundation of all mathematical knowledge must be laid, in definitions and selfevident truths. A *definition* is an explanation of what is meant, by any word or phrase. Thus, an equilateral triangle is defined, by saying, that it is a figure bounded by three equal sides.

It is essential to a complete definition, that it perfectly distinguish the thing defined, from every thing else. On many subjects, it is difficult to give such precision to language, that it shall convey, to every hearer or reader, exactly the same ideas. But, in the mathematics, the principal terms may be so defined, as not to leave room for the least difference of apprehension, respecting their meaning. All must be agreed, as to the nature of a circle, a square, and a triangle, when they have once learned the definitions of these figures.

Under the head of definitions, may be included explanations of the *characters* which are used to denote the relations of quantities. Thus, the character $\sqrt{}$ is explained or defined, by saying that it signifies the same as the words square root.

10. The next step, after becoming acquainted with the meaning of mathematical terms, is to bring them together, in the form of propositions. Some of the relations of quantities require no process of reasoning, to render them evident. To be understood, they need only to be proposed. That a square is a different figure from a circle; that the whole of a thing is greater, than one of its parts; and, that two straight lines cannot inclose a space, are propositions so manifestly true, that no reasoning upon them could make them more certain. They are, therefore, called selfevident truths, or *axioms*.

11. There are, however, comparatively few mathematical truths which are selfevident. Most require to be proved, by a chain of reasoning. Propositions of this nature are denominated *theorems;* and the process, by which, they are shewn to be true, is called

demonstration. This is a mode of arguing, in which, every inference is immediately derived, either from definitions and selfevident axioms, or from principles which have been previously demonstrated. In this way, complete certainty is made to accompany every step, in a long course of reasoning.

12. Demonstration is either *direct*, or *indirect*. The former is the common, obvious mode of conducting a demonstrative argument. But, in some instances, it is necessary to resort to indirect demonstration; which is a method of establishing a proposition, by proving that to suppose it *not* true, would lead to an absurdity. This is frequently called *reductio ad absurdum.* Thus, in certain cases in geometry, two lines may be proved to be equal, by shewing that to suppose them unequal, would involve an absurdity.

13. Besides the principal theorems in the mathematics, there are also *Lemmas*, and *Corollaries.* A Lemma is a proposition which is demonstrated, for the purpose of using it, in the demonstration of some other proposition. This preparatory step is taken, to prevent the proof of the principal theorem from becoming complicated and tedious.

14. A *Corollary* is an *inference* from a preceding proposition. A Scholium is a remark of any kind, suggested by something which has gone before, though not, like a corollary, immediately depending on it.

15. The immediate object of inquiry, in the mathematics, is, frequently, not the demonstration of a general truth, but a method of performing some operation, such as reducing a vulgar fraction to a decimal, extracting the cube root, or inscribing a circle in a square. This is called solving a problem. A *theorem* is something to be *proved.* A *problem* is something to be *done.*

16. When that which is required to be done, is so easy, as to be obvious to every one, without an explanation, it is called a *postulate.* Of this nature, is

the drawing of a straight line, from one point to another. A postulate is, to a problem, what an axiom is, to a theorem.

17. A quantity is said to be *given*, when it is either supposed to be already *known*, or is made a *condition*, in the statement of any theorem or problem. In the rule of proportion in arithmetic, for instance, three terms must be given, to enable us to find a fourth.— These three terms are the *data*, upon which the calculation is founded. If we are required to find the number of acres, in a circular island ten miles in circumference, the circular figure, and the length of the circumference, are the data. They are said to be given *by supposition*, that is, by the conditions of the problem. A quantity is also said to be given, when it may be directly and easily *inferred*, from something else which is given. Thus, if two numbers are given, their *sum* is given; because it is obtained, by merely adding the numbers together.

In Geometry, a quantity may be given, either in *position*, or magnitude, or both. A line is given in position, when its *situation* and *direction* are known.— It is given in magnitude, when its *length* is known. A circle is given in *position*, when the place of its centre is known. It is given in *magnitude*, when the length of its diameter is known.

18. One quantity is *contrary*, or contradictory to another, when what is affirmed, in the one, is denied, in the other. A proposition and its contrary, can never *both* be true. It cannot be true, that two given lines are equal, and that they are *not* equal, at the same time.

19. One proposition is the *converse* of another, when the order is inverted; so that, what is *given* or supposed, in the first, becomes the *conclusion*, in the last; and what is given in the last, is the conclusion, in the first. Thus, it can be proved, first, that if the *sides* of a triangle are equal, the *angles* are equal; and secondly, that if the *angles* are equal, the *sides*

are equal. Here, in the first proposition, the equality of the *sides is given;* and the equality of the *angles*,
inferred: in the second, the equality of the *angles* is
given, and the equality of the sides inferred. In many instances, a proposition and its converse are both
true; as in the preceding example. But this is not
always the case. A circle is a figure bounded by a
curve; but a figure bounded by a curve is not of
course a circle.

20. The practical applications of the mathematics,
in the common concerns of business, in the useful
arts, and in the various branches of physical science,
are almost innumerable. Mathematical principles are
necessary, in *mercantile transactions*, for keeping, arranging, and settling accounts, adjusting the prices of
commodities, and calculating the profits of trade : in
Navigation, for directing the course of a ship on the
ocean, adapting the position of her sails to the direction of the wind, finding her latitude and longitude,
and determining the bearings and distances of objects
on shore : in *Surveying*, for measuring, dividing, and
laying out grounds, taking the elevation of hills, and
fixing the boundaries of fields, estates and public ter-
ritories : in *Mechanics*, for understanding the laws of
motion, the composition of forces, the equilibrium of
the mechanical powers, and the structure of machines:
in *Architecture*, for calculating the comparative strength
of timbers, the pressure which each will be required to
sustain, the forms of arches, the proportions of col-
umns, &c. : in *Fortification*, for adjusting the position,
lines, and angles, of the several parts of the works :
in *Gunnery*, for regulating the elevation of the can-
non, the force of the powder, and the velocity and
range of the shot : in *Optics*, for tracing the direction
of the rays of light, understanding the formation of
images, the laws of vision, the separation of colours,
the nature of the rainbow, and the construction of mi-
croscopes and telescopes : in *Astronomy*, for computing the distances, magnitudes, and revolutions of the

heavenly bodies; and the influence of the law of grav-
itation, in raising the tides, disturbing the motions of
the moon, causing the return of the comets, and retaining the planets in their orbits : in *Geography*,
for determining the figure and dimensions of the
earth, the extent of oceans, islands, continents, and
countries ; the latitude and longitude of places, the
courses of rivers, the height of mountains, and the
boundaries of kingdoms : in *History*, for fixing the
chronology of remarkable events, and estimating the
strength of armies, the wealth of nations, the value of
their revenues, and the amount of their population :
and, in the concerns of *Government*, for apportioning
taxes, arranging schemes of finance, and regulating
national expenses. The mathematics have also important applications to Chemistry, Mineralogy, Music,
Painting, Sculpture, and indeed to a great proportion
of the whole circle of arts and sciences.

21. It is true, that, in many of the branches which
have been mentioned, the ordinary business is frequently transacted, and the mechanical operations
performed, by persons who have not been regularly
instructed in a course of mathematics. Machines are
framed, lands are surveyed, and ships are steered, by
men who have never thoroughly investigated the principles, which lie at the foundation of their respective
arts. The reason of this is, that the methods of proceeding, in their several occupations, have been point-
ed out to them, by the genius and labour of others.
The mechanic often works by rules, which men of
science have provided, for his use, and of which he
knows nothing more, than the practical application.
The mariner calculates his longitude by tables, for
which he is indebted to mathematicians and astrono-
mers of no ordinary attainments. In this manner,
even the abstruse parts of the mathematics are made
to contribute their aid, to the common arts of life.

22. But an additional and more important advantage, to persons of a liberal education, is to be found,

in the enlargement and improvement of the reasoning powers. The mind, like the body, acquires strength by exertion. The art of reasoning, like other arts, is learned by practice. It is perfected, only by long continued exercise. Mathematical studies are peculiarly fitted for this discipline of the mind. They are calculated to form it to habits of fixed attention; of sagacity, in detecting sophistry; of caution, in the admission of proof; of dexterity, in the arrangement of arguments; and of skill, in making all the parts of a long continued process tend to a result, in which the truth is clearly and firmly established. When a habit of close and accurate thinking is thus acquired; it may be applied to any subject, on which a man of letters or of business may be called to employ his talents. "The youth," says Plato, "who are furnished with mathematical knowledge, are prompt and quick, at all other sciences."

It is not pretended, that an attention to other objects of inquiry, is rendered unnecessary, by the study of the mathematics. It is not their office, to lay before us historical facts; to teach the principles of morals; to store the fancy with brilliant images; or to enable us to speak and write with rhetorical vigour and elegance. The beneficial effects which they produce on the mind, are to be seen, principally, in the regulation and increased energy of the *reasoning powers*. These they are calculated to call into frequent and vigorous exercise. At the same time, mathematical studies may be so conducted, as not often to require excessive exertion and fatigue. Beginning with the more simple subjects, and ascending gradually to those which are more complicated; the mind acquires strength, as it advances; and by a succession of steps, rising regularly one above another, is enabled to surmount the obstacles which lie in its way. In a course of mathematics, the parts succeed each other in such a connected series, that the preceding propositions are preparatory to those which follow.

The student who has made himself master of the former, is qualified for a successful investigation of the latter. But he who has passed over any of the ground superficially, will find that the obstructions to his future progress are yet to be removed. In mathematics, as in war, it should be made a principle, not to advance, while any thing is left unconquered behind. It is important that the student should be deeply impressed with a conviction of the necessity of this. Neither is it sufficient that he understands the *nature* of one proposition or method of operation, before proceeding to another. He ought also to make himself *familiar* with every step, by a careful attention to the examples. He must not expect to become thoroughly versed in the science, by merely *reading* the main principles, rules and observations. It is practice only, which can put these completely in his possession. The method of studying, here recommended, is not only that which promises success, but that which will be found, in the end, to be the most expeditious, and by far the most pleasant. While a superficial attention occasions perplexity and consequent aversion ; a thorough investigation is rewarded with a high degree of gratification. The peculiar entertainment which mathematical studies are calculated to furnish to the mind, is reserved for those who make themselves masters of the subjects to which their attention is called.

NOTE. The principal definitions, theorems, rules, &c. which it is necessary to *commit to memory*, are distinguished by being put in Italics.

C

ALGEBRA.

SECTION I.

Notation, Negative Quantities, Axioms, &c.

ART. 23. ALGEBRA may be defined, *a general method of investigating the relations of quantities, principally by letters.* This, it must be acknowledged, is an imperfect account of the subject; as every account must necessarily be, which is comprised in the compass of a definition. Its real nature is to be learned, rather by an attentive examination of its parts, than from any summary description.

The solutions in Algebra, are of a more *general* nature, than those in common Arithmetic. The latter relate to particular numbers; the former, to whole *classes* of quantities. On this account, Algebra has been termed a kind of *universal Arithmetic.* The generality of its solutions is principally owing to the use of *letters*, instead of numeral figures, to express the several quantities which are subjected to calculation. In Arithmetic, when a problem is solved, the answer is limited to the particular numbers which are specified, in the statement of the question. But an algebraic solution may be equally applicable to all other quantities which have the same relations. This important advantage is owing to the difference between the customary use of figures, and the manner in which letters are employed in Algebra. One of the nine digits invariably expresses the same number: but a letter may be put for any number whatever. The figure 8 always signifies eight; the figure 5, five, &c. And, though one of the digits, in connection with others, may have a *local* value, different from its simple value when alone; yet the same *combination* always expresses the same number. Thus 263 has one uniform signification. And this is the case with every other combination of figures. But in Algebra, a letter may stand for any quantity which we wish it to represent. Thus *b* may be put for 2, or 10, or 50, or 1000. It must not be understood from this, however, that the letter has no determinate value. Its value is fixed for the occasion. For the present purpose, it remains unaltered. But on a different occasion, the same letter may be put for any other number.

A calculation may be greatly abridged by the use of letters; especially when very large numbers are concerned. And when several such numbers are to be combined, as in multiplication, the process becomes extremely tedious. But a single letter may be put for a large number, as well as for a small one. The numbers 26347297, 68347823, and 27462498, for instance, may be expressed by the letters *b, c* and *d*. The multiplying them together, as will be seen hereafter, will be nothing more, than writing them, one after another, in the form of a word, and the product will be simply *bcd*. Thus, in Algebra, much of the labour of calculation may be saved, by the rapidity of the operations. Solutions are sometimes effected, in the compass of a few lines, which, in common Arithmetic, must be extended through many pages.

24. Another advantage obtained from the notation by letters instead of figures, is, that the several quantities which are brought into a calculation, may be preserved *distinct from each other*, though carried through a number of complicated processes; whereas, in arithmetic, they are so blended together, that no trace is left of what they were, before the operation began. To give a very simple example: suppose it is required to multiply together 74, 23 and 41. By arithmetic, the product is found to be 69782. This product, however, would not of itself suggest

the numbers which had been multiplied together to
produce it. But in algebra, if 74 be represented by
a, 23, by b, and 41, by c; the product will be abc,
which not only stands for 69782; but also indicates
the factors, a, b, and c, which were multiplied, to obtain this product.

25. Algebra differs farther from arithmetic, in
making use of *unknown* quantities, in carrying on its
operations. In arithmetic, all the quantities which
enter into a calculation must be known. For they
are expressed *in numbers*. And every number must
necessarily be a determinate quantity. But in algebra, a letter may be put for a quantity, before its
value has been ascertained. And yet it may have
such relations, to other quantities, with which it is
connected, as to answer an important purpose in the
calculation.

NOTATION.

26. To facilitate the investigations in algebra, the
several steps of the reasoning, instead of being expressed in *words*, are translated into the language of
signs and symbols, which may be considered as a
species of *short-hand*. This serves to place the quantities and their relations distinctly before the eye, and
to bring them all into view at once. They are thus
more readily compared and understood, than when
removed at a distance from each other, as in the
common mode of writing. But before any one can
avail himself of this advantage, he must become perfectly familiar with the new language.

27. The *quantities* in algebra, as has been already
observed, are generally expressed by *letters*. The
first letters of the alphabet are used, to represent
known quantities; and the *last* letters, those which are
unknown. Thus b may be put for a known, and y, for
an unknown quantity. Sometimes the quantities, in-

stead of being expressed by letters, are set down in figures, as in common arithmetic.

28. Besides the letters and figures, there are certain characters used, to indicate the *relations* of the quantities, or the *operations* which are performed with them. Among these are the signs + and —, which are read *plus* and *minus*, or *more* and *less*. The former is prefixed to quantities which are to be *added*; the latter, to those which are to be *subtracted*. Thus $a + b$ signifies that b is to be added to a. If a stands for 10, and b, for 6; then $a + b$ is 16. It is read a plus b, or a added to b, or a and b. If the expression be $a - b$, i. e. a minus b; it indicates that b is to be subtracted from a. In figures $10 - 6$ is 10 diminished by 6 i. e. 4.

29. The sign + is prefixed to quantities which are considered as *affirmative* or *positive*; and the sign —, to those which are supposed to be negative. For the nature of this distinction, see art. 54.

All the quantities which enter into an algebraic process, are considered, for the purposes of calculation, as either positive or negative. Before the *first* one, unless it be negative, the sign is generally omitted. But it is always to be understood. Thus $a + b$, is the same as $+ a + b$. For a is as much added to b, as b is, to a. They are added *together:* $a + b$ is a and b; or, which is the same thing, b and a.

30. Sometimes *both* + and — are prefixed to the same letter. The sign is then said to be *ambiguous*. Thus $a \pm b$ signifies that in certain cases, comprehended in a general solution, b is to be added to a, and, in other cases, subtracted from it.

31. When it is intended to express the difference between two quantities without deciding which is the one to be subtracted, the character ∞ or $\backsim$ is used. Thus $a \backsim b$, or $a \infty b$ denotes the difference between a and b, without determining whether a is to be subtracted from b, or b from a.

32. The *equality* between two quantities or sets of quantities is expressed, by parallel lines =. Thus $a + b = d$ signifies that a and b together are equal to d. So $8 + 3 = 11$ i. e. 8 and 3 equal 11. And $a + d = c = b + g = h$ signifies that a and d equal c, which is equal to b and g, which are equal to h.— Again $8 + 4 = 16 - 4 = 10 + 2 = 7 + 2 + 3 = 12$. This is read $8 + 4$ is equal to $16 - 4$, which is equal to $10 + 2$, which is equal to $7 + 2 + 3$, which is equal to 12.

33. When the first of the two quantities compared is *greater*, than the other, the character $>$ is placed between them. Thus $a > b$ signifies that a is greater than b.

If the first is *less* than the other, the character $<$ is used; as $a < b$; i. e. a is less than b. In both cases, the quantity towards which the character *opens*, is greater than the other.

34. A numeral figure is often prefixed to a letter. This is called a *co-efficient*. It shows how often the quantity expressed by the letter is to be taken. Thus $2b$ signifies twice b, and $9b$, 9 times b, or 9 multiplied into b. If b stands for 10, then $9b$ is 9 times 10 or 90.

The co-efficient may be either a whole number or a fraction. Thus $\frac{2}{3}b$ is two thirds of b. When the co-efficient is not expressed, 1 is always to be understood. Thus a is the same as $1a$ i. e. once a.

35. The co-efficient may be a *letter*, as well as a figure. In the quantity mb, m may be considered the co-efficient of b; because b is to be taken as many times as there are units in m. If b stands for 6, then mb is 6 times b. In $3abc$, 3 may be considered as the co-efficient of abc; $3a$, the co-efficient of bc; or $3ab$, the co-efficient of c. See art. 42.

36. A *simple* quantity is either a single letter or number, or several letters connected together, without the signs $+$ and $-$. Thus a, ab, abd, and $8b$ are each of them simple quantities. A *compound*

quantity consists of a number of simple quantities, connected by the sign + or —. Thus $a + b$, $d — y$, $b — d + 3h$, are each compound quantities. The members, of which it is composed, are called *terms*.

37. If there are *two* terms in a compound quantity, it is called a *binomial*. Thus $a + b$ and $a — b$ are binomials. The latter is also called a *residual* quantity, because it expresses the difference of two quantities, or the remainder, after one is taken from the other. A compound quantity consisting of *three* terms, is sometimes called a *trinomial*; one of *four* terms, a *quadrinomial*, &c.

38. When the several members of a compound quantity are to be subjected to the same operation, they are frequently connected by a line called a *vinculum*. Thus $a — \overline{b + c}$ shows that the *sum* of b and c is to be subtracted from a. But $a — b + c$ signifies that b only is to be subtracted from a, while c is to be added. The sum of c and d, subtracted from the sum of a and b, is $\overline{a + b} — \overline{c + d}$. The marks used for parentheses, ( ) are often substituted, instead of a line, for a vinculum. Thus $x — (a + c)$ is the same as $x — \overline{a + c}$. The *equality* of two sets of quantities is expressed, without using a vinculum.— Thus $a + b = c + d$ signifies, not that b is equal to e; but that the sum of a and b is equal to the sum of c and d.

39. A single letter, or a number of letters, representing any quantities with their relations, is called an *algebraic expression*; and sometimes a *formula*. Thus $a + b + 3d$ is an algebraic expression.

40. The character × denotes *multiplication*. Thus $a × b$ is a multiplied into b: and $6 × 3$ is 6 times 3, or 6 into 3. Sometimes a *point* is used to indicate multiplication. Thus $a . b$ is the same as $a × b$. But the sign of multiplication is more commonly omitted, between simple quantities; and the letters are connected together, in the form of a word or syllable.—

Thus ab is the same as $a.b$ or $a \times b$. And $bcde$ is the same as $b \times c \times d \times e$. When a compound quantity is to be multiplied, a *vinculum* is used, as in the case of subtraction. Thus the sum of a and b, multiplied into the sum of c and d, is $\overline{a + b} \times \overline{c + d}$, or $(a + b) \times (c + d)$. And $(6 + 2) \times 5$ is 8×5, or 40. But $6 + 2 \times 5$ is $6 + 10$ or 16. When the marks of parenthesis are used, the sign of multiplication is frequently omitted. Thus $(x + y) (x - y)$ is $(x + y) \times (x - y)$.

41. When two or more quantities are multiplied together, each of them is called a *factor*. In the product ab, a is a factor, and so is b. In the product $x \times \overline{a + m}$, x is one of the factors, and $a + m$, the other. Hence every *co-efficient* may be considered a factor. (Art. 35.) In the product $3y$, 3 is a factor, as well as y.

42. A quantity is said to be *resolved into factors*, when any factors are taken which, being multiplied together, will produce the given quantity. Thus $3ab$ may be resolved into the two factors $3a$ and b because $3a \times b$ is $3ab$. And $5amn$ may be resolved into the three factors $5a$, and m, and n; because $5a \times m \times n$ is $5amn$. And 48 may be resolved, into the two factors 2×24, or 3×16, or 4×12, or 6×8; or into the three factors $2 \times 3 \times 8$, or $4 \times 6 \times 2$, &c.

43. The character $\div$ is used to show, that the quantity which precedes it, is to be *divided*, by that which follows. Thus $a \div c$ is a divided by c: and $\overline{a + b} \div \overline{c + d}$ is the sum of a and b, divided by the sum of c and d. But in algebra, division is more commonly expressed, by writing the divisor under the dividend, in the form of a vulgar fraction. Thus $\frac{a}{b}$ is the same as $a \div b$: and $\frac{c - b}{d + h}$ is the difference of c and b divided by the sum of d and h. A character prefixed to the dividing line of a fractional ex-

pression, is to bé understood as referring to all the parts taken collectively; that is, to the whole value of the quotient. Thus $a - \dfrac{b + c}{m + n}$ signifies that the quotient of $b + c$ divided by $m + n$ is to be subtracted from a. And $\dfrac{c - {}'d}{a + m} \times \dfrac{h + n}{x - y}$ denotes that the first quotient is to be multiplied into the second.

44. When four quantities are *proportional*, the proportion is expressed by points, in the same manner, as in the Rule of Three in arithmetic. Thus $a : b :: c : d$ signifies that a has to b, the same ratio, which c has to d. And $ab : cd :: a + m : b + n$, means, that ab is to cd; as the sum of a and m, to the sum of b and n.

45. Algebraic quantities are said to be *alike*, when they are expressed by the same *letters*, and are of the same *power:* and *unlike*, when the letters are different, or when the same letter is raised to different powers.* Thus $ab, 3ab, -ah$, and $-6ab$, are like quantities, because the letters are the same in each, although the signs and co-efficients are different. But $3a, 3y$, and $3bx$, are unlike quantities, because the letters are unlike, although there is no difference in the signs and co-efficients.

46. One quantity is said to be a *multiple* of another, when the former *contains* the latter a certain number of times, without a remainder. Thus $10a$ is a multiple of $2a$, because the one contains the other just 5 times. So 24 is a multiple of 6.

47. One quantity is said to be a *measure* of another, when the former *is contained* in the latter, any number of times, without a remainder. Thus $3b$ is a measure of $15b$: and 7 is a measure of 35.

48. The *value* of an expression, is the number or

---

* For the notation of *powers* and *roots*, see the sections on those subjects.

D

quantity, for which the expression stands. Thus the value of $3 + 4$ is 7; of 3×4 is 12; of $\frac{16}{8}$ is 2.

49. *The* RECIPROCAL *of a quantity, is the quotient arising from dividing an unit by that quantity.* Thus the reciprocal of a is $\frac{1}{a}$; the reciprocal of $a + b$ is $\frac{1}{a + b}$; the reciprocal of 4 is $\frac{1}{4}$.

50. The relations of quantities, which, in ordinary language, are signified by *words*, are represented, in the algebraic notation, by *signs*. The latter mode of expressing these relations, ought to be made so familiar to the mathematical student, that he can, at any time, substitute the one for the other. A few examples are here added, in which, words are to be converted into signs.

1. What is the algebraic expression for the following statement, in which, the letters a, b, c, &c. may be supposed to represent any given quantities?

The product of $a, b,$ and $c,$ divided by the difference of c and d, is equal to the sum of b and c added to 15 times h.

Ans. $\dfrac{abc}{c - d} = b + c + 15h.$

2. The product of the difference of a and h into the sum of $b, c,$ and d, is equal to 37 times m, added to the quotient of b divided by the sum of h and b.— Ans.

3. The sum of a and b, is to the quotient of b divided by c; as the product of a into c, to 12 times h. Ans.

4. The sum of $a, b,$ and c divided by six times their product, is equal to four times their sum diminished by d. Ans.

5. The quotient of 6 divided by the sum of a and b, is equal to 7 times d, diminished by the quotient of b, divided by 36. Ans.

51. It is necessary also, to be able to reverse what

is done in the preceding examples, that is, to translate the algebraic signs into common language.

What will the following expressions become, when
words are substituted for the signs?

1. $\dfrac{a + b}{h} = a\,b\,c - 6\,m + \dfrac{a}{a + c}.$

Ans. The sum of a and b divided by h, is equal to
the product of a, b, and c, diminished by 6 times m,
and increased by the quotient of a divided by the
sum of a and c.

2. $a\,b + \dfrac{3\,h - c}{x + y} = d \times \overline{a + b + c} - \dfrac{h}{6 + b}.$

Ans.

3. $a + 7\,(h + x) - \dfrac{c - 6\,d}{2\,a + 4} = (a + h) \times (b - c).$

Ans.

4. $a - b : a\,c :: \dfrac{d - 4}{m} : 3 \times \overline{h + d + y}.$

Ans.

5. $\dfrac{a - h}{3 + \overline{b - c}} + \dfrac{d + a\,b}{2\,m} = \dfrac{b\,a \times \overline{d + h}}{a\,m} - \dfrac{c\,d}{h + d\,m}.$

Ans.

52. At the close of an algebraic process, it is frequently necessary to restore the *numbers*, for which
letters had been substituted, at the beginning. In doing this, the sign of multiplication must not be omitted, as it generally is, between factors expressed by
letters. Thus if a stands for 3, and b, for four; the
product $a.b$ is not 34, but 3×4 i. e. 12.

In the following examples,

Let $a = 3$ And $d = 6$.
    $b = 4$ $m = 8$.
    $c = 2$ $n = 10$.

Then, 1. $\dfrac{a + m}{c\,d} + \dfrac{b\,c - n}{3\,d} = \dfrac{3 + 8}{2 \times 6} + \dfrac{4 \times 2 - 10}{3 \times 6}$.

2. $\dfrac{b + a\,d}{c - d\,m} - b\,c\,m\,n + \dfrac{d - 4\,c\,n}{5\,a\,b} = \underline{\quad\quad}$

3. $b\,m\,d + \dfrac{a\,b - 3\,d}{c\,d\,m} - \dfrac{3\,b\,n - b\,c}{4\,a + 3\,c\,d} + \dfrac{b}{a} = \underline{\quad}$

53. An algebraic expression, in which numbers have, in this manner, been substituted for letters, may often be rendered much more simple, by reducing several terms to one. This can not generally be done, while the letters remain. If $a + b$ is used for the sum of two quantities, a can not be united in the same term with b. But if a stands for 3, and b, for 4, then $a + b = 3 | 4 = 7$. The value of an expression consisting of many terms may thus be found, by actually performing, with the numbers, the operations of addition, subtraction, multiplication, &c. indicated by the algebraic characters.

Find the value of the following expressions, in which, the letters are supposed to stand for the same numbers, as in the preceding article.

1. $\dfrac{a\,d}{c} + a + m\,n = \dfrac{3 \times 6}{2} + 3 + 8 \times 10 = 9$
$+ 3 + 80 = 92$.

2. $a\,b\,m + \dfrac{2\,b}{m - d} + 2n = 3 \times 4 \times 8 + \dfrac{2 \times 4}{8 - 6}$
$+ 2 \times 10 = 120$.

3. $\overline{a + c} \times \overline{n - m} + \dfrac{m - b}{m - d} - a \times \overline{n - m} = 6$.

$$4. \quad \frac{a \times \overline{d+c}}{n-d} + abc - \frac{\overline{c+b} \times \overline{m-d}}{n-bc} = \underline{\quad}$$

$$5. \quad \frac{ac+5m}{2n+3} + m - cb + \frac{\overline{4d-b} \times \overline{a-c}}{n} = $$

POSITIVE and NEGATIVE QUANTITIES.*

54. To one who has just entered on the study of algebra, there is generally nothing more perplexing, than the use of what are called *negative* quantities.— He supposes he is about to be introduced, to a class of quantities, which are entirely new; a sort of mathematical *nothings*, of which he can form no distinct conception. As positive quantities are *real*, he concludes that those which are negative must be *imaginary*. But this is owing to a misapprehension of the term negative, as used in the mathematics.

55. A NEGATIVE *quantity is one which is required to be* SUBTRACTED. When several quantities enter into a calculation, it is frequently necessary that some of them should be *added* together, while others are *subtracted*. The former are called affirmative or positive, and are marked with the sign +; the latter are termed negative, and distinguished by the sign —. If, for instance, the profits of trade are the subject of calculation, and the *gain* is considered positive; the *loss* will be negative; because the latter must be *subtracted* from the former, to determine the clear profit. If the sums of a book account, are brought into an algebraic process, the debt and the credit are distinguished by opposite signs. If a man on a journey is, by any accident necessitated to return several miles,

* On the subject of Negative quantities, see Newton's Universal Arithmetic, Maseres on the Negative Sign, Mansfield's Mathematical Essays,‡ and Maclaurin's, Simpson's, Euler's, Saunderson's and Ludlam's Algebra.

this backward motion is to be considered *negative,* because that, in determining his real progress, it must be subtracted, from the distance which he has travelled in the opposite direction. If the *ascent* of a body from the earth be called positive, its *descent* will be negative. These are only different examples of the same general principle. In each of the instances, one of the quantities is to be *subtracted* from the other.

56. The terms positive and negative, as used in the mathematics, are merely *relative.* They imply that there is, either in the nature of the quantities, or in their circumstances, or in the purposes which they are to answer in calculation, some such *opposition* as requires that one should be *subtracted* from the other. But this opposition is not that of existence and non-existence, nor of one thing greater than nothing, and another less than nothing. For, in many cases, either of the signs may be, indifferently and at pleasure, applied to the very same quantity; that is, the two characters may change places. In determining the progress of a ship, for instance, her easting may be marked +, and her westing −; or the westing may be +, and the easting −. All that is necessary is, that the two signs be prefixed to the quantities, in such a manner as to show, which are to be added, and which subtracted. In different processes, they may be differently applied. On one occasion, a downward motion may be called positive, and on another occasion, negative.

57. In every algebraic calculation, some one of the quantities must be fixed upon, to be considered positive. All other quantities which will *increase* this, must be positive also. But those which will tend to *diminish* it, must be negative. In a mercantile concern, if the *stock* is supposed to be positive, the *profits* will be positive; for they *increase* the stock; they are to be *added* to it. But the *losses* will be negative; for they *diminish* the stock; they are to be *subtracted*

from it. When a boat, in attempting to ascend a river, is occasionally driven back by the current; if the progress up the stream, to any particular point, is considered positive, every succeeding instance of *forward* motion will be positive, while the *backward* motion will be negative.

58. A negative quantity is frequently *greater*, than the positive one with which it is connected. But how, it may be asked, can the former be *subtracted* from the latter? The greater is certainly not *contained* in the less: how then can it be taken out of it? The answer to this is, that the greater may be supposed first to *exhaust* the less, and then to leave a remainder equal to the difference between the two. If a man has in his possession, 1000 dollars, and has contracted a debt of 1500; the latter subtracted from the former, not only exhausts the whole of it, but leaves a balance of 500 against him. In common language, he is 500 dollars worse than nothing.

59. In this way, it frequently happens, in the course of an algebraic process, that a negative quantity is brought to *stand alone*. It has the sign of subtraction, without being connected with any other quantity, from which it is to be subtracted. This denotes that a previous subtraction has left a remainder, which is a part of the quantity subtracted. If the latitude of a ship which is 20 degrees north of the equator, is considered positive, and if she sails south 25 degrees; her motion first *diminishes* her latitude, then reduces it to *nothing*, and finally gives her 5 degrees of *south* latitude. The sign —, prefixed to the 25 degrees is retained before the 5, to show that this is what remains of the *southward* motion, after balancing the 20 degrees of north latitude. If the motion southward is only 15 degrees, the remainder must be + 5, instead of − 5, to show that it is a part of the ship's *northern* latitude, which has been thus far diminished, but not reduced to nothing. The balance of a book account will be positive or negative, according as the

debt or the credit is the greater of the two. To determine to which side the remainder belongs, the sign
must be retained, though there is no other quantity,
from which this is again to be subtracted, or to which
it is to be added.

60. When a quantity continually decreasing is reduced to nothing, it is sometimes said to become afterwards *less than nothing*. But this is an exceptionable manner of speaking.* No quantity can be really less than nothing. It may be diminished, till it
vanishes, and gives place to an *opposite* quantity. The
latitude of a ship crossing the equator, is first made
less, then nothing, and afterwards *contrary* to what it
was before. The north and south latitudes may
therefore be properly distinguished, by the signs +
and — ; all the positive degrees being on one side of
0, and all the negative, on the other; thus,

$+6, +5, +4, +3, +1, 0, -1, -2, -3, -4, -5, \&c.$

The numbers belonging to any other series of opposite quantities, may be arranged in a similar manner. So that 0 may be conceived to be a kind of
dividing point between positive and negative numbers.
On a thermometer, the degrees *above* 0 may be considered positive, and those *below* 0, negative.

61. A quantity is sometimes said to be *subtracted
from* 0. By this is meant, that it belongs on the negative side of 0. But a quantity is said to be *added* to 0,
when it belongs on the positive side. Thus in speaking of the degrees of a thermometer, $0 + 6$ means 6
degrees *above* 0; and $0 - 6$, 6 degrees *below* 0.

* NOTE. The expression "*less than nothing*" may not be
wholly improper; if it is intended to be understood, not literally, but merely as a convenient phrase adopted for the sake
of avoiding a tedious circumlocution; as we say "the sun ri-
ses," instead of saying "the earth rolls round, and brings the
sun into view." The use of it in this manner, is warranted by
Newton, Euler, and others.

AXIOMS.

62. The object of mathematical inquiry is, generally, to investigate some unknown quantity, and discover how *great* it is. This is effected, by comparing it with some other quantity or quantities already known. The dimensions of a stick of timber are found, by applying to it a measuring rule of known length. The *weight* of a body is ascertained, by placing it in one scale of a balance, and observing how many pounds in the opposite scale, will equal it. And any quantity is determined, when it is found to be equal to some known quantity or quantities.

Let a and b be known quantities, and y, one which is unknown. Then y will become known, if it is discovered to be equal to the sum of a and b: that is, if

$$y = a + b.$$

An expression like this, representing the equality between one quantity or set of quantities, and another, is called an *equation*. It will be seen hereafter, that much of the business of algebra consists in finding equations, in which, some unknown quantity is shown to be equal to others which are known. But it is not often the fact, that the first comparison of the quantities, furnishes the equation required. It will generally be necessary to make a number of additions, subtractions, multiplications, &c. before the unknown quantity is discovered. But in all these changes, a constant equality must be preserved, between the two sets of quantities compared. This will be done, if in making the alterations, we are guided by the following *axioms.* These are not inserted here, for the purpose of being proved; for they are self-evident. (Art. 10.) But as they must be continually introduced or implied, in demonstrations and the solutions of problems, they are placed together, for the convenience of reference.

63. Axiom 1. If the same quantity or equal quantities be *added* to equal quantities, their *sums* will be equal.

2. If the same quantity or equal quantities be *subtracted* from equal quantities, the *remainders* will be equal.

3. If equal quantities be *multiplied* into the same, or equal quantities, the *products* will be equal.

4. If equal quantities be *divided* by the same or equal quantities, the *quotients* will be equal.

E

5. If the same quantity be both *added to* and *subtracted from* another, the value of the latter will not be altered.

6. If a quantity be both *multiplied* and *divided* by another, the value of the former will not be altered.

7. If to unequal quantities, equals be added, the greater will give the greater sum.

8. If from unequal quantities, equals be subtracted, the greater will give the greater remainder.

9. If unequal quantities be multipled by equals, the greater will give the greater product.

10. If unequal quantities be divided by equals, the greater will give the greater quotient.

11. Quantities which are respectively equal to any other quantity, are equal to each other.

12. The whole of a quantity is greater than a part.

This is, by no means a *complete* list of the self-evident propositions, which are furnished by the mathematics. It is not necessary to enumerate them all. Those have been selected, to which we shall have the most frequent occasion to refer.

$$\text{If } a = bc$$

Then by ax.1. $a + m = bc + m$

ax.2. $a - m = bc - m$

ax.3. $am = bcm$

ax.4. $\dfrac{a}{m} = \dfrac{bc}{m}$

ax.5. $a = a + m - m$

ax.6. $a = \dfrac{am}{m}$

And if $a > cd$

By ax.7. $\overline{a + m} > \overline{cd + m}$

ax.8. $a - m > cd - m$

ax.9. $am > cdm$

ax.10. $\dfrac{a}{m} > \dfrac{cd}{m}$

If $b = ch$

And $d = ch$

By ax.11. $b = d$.

64. The investigations in algebra are carried on principally, by means of a series of *equations* and *proportions*. But instead of entering directly upon these, it will be necessary to attend, in the first place, to a number of processes, on which the management of equations and proportions depends. These preparatory operations are similar to the calculations under the common rules of arithmetic. We have addition, multiplication, division, involution, &c. in algebra, as well as in arithmetic. But this application of a common name, to operations in these two branches of the mathemat-

ics, is often the occasion of perplexity and mistake. The learner naturally expects to find addition in algebra the same as addition in arithmetic. They are in fact the same, in many respects: in *all* respects perhaps, in which the steps of the one will admit of a direct comparison, with those of the other. But addition in algebra is more *extensive*, than in arithmetic. The same observation may be made, concerning several other operations in algebra. They are, in many points of view, the same as those which bear the same names in arithmetic. But they are frequently extended farther, and comprehend processes which are unknown to arithmetic. This is commonly owing to the introduction of negative quantities. The management of these requires steps which are unnecessary, where quantities of one class only are concerned. It will be important therefore, as we pass along, to mark the *difference*, as well as the *resemblance*, between arithmetic and algebra; and, in some instances, to give a new definition, accommodated to the latter.

SECTION II.

ADDITION.

ART. 65. IN entering on an algebraic calculation, the first thing to be done, is evidently to *collect the materials*. Several distinct quantities are to be concerned in the process. These must be brought together. They must be connected in some form of expression, which will present them at once to our view, and show the relations which they have to each other. This collecting of quantities is what, in algebra, is called *addition*. It may be defined, *the connecting of several quantities, with their signs, in one algebraic expression*.

66. It is common to include in the definition, "uniting in one term, such quantities, as will admit of being united." But this is not so much a part of the addition itself, as a *reduction*, which accompanies or follows it. The addition may, in all cases, be performed, by merely connecting the quantities, by their proper signs. Thus a added to b is, evidently, a and b : that is, according to the algebraic notation $a+b$. And a, added to the sum of b and c, is $a+b+c$.——— And $a+b$, added to $c+d$, is $a+b+c+d$. In the same manner, if the sum of any quantities whatever, be added to the sum of any others, the expression for the whole, will contain all these quantities, connected by the sign $+$.

Thus the sum of $a+2b$ and $4c+d+h$ and $m+y$, is

$$a+2b+4c+d+h+m+y.$$

67. Again, if the *difference* of a and b be added to c; the sum will be $a-b$ added to c, that is $a-b+c$. And if $a-b$ be added to $c-d$, the sum will be $a-b+c-d$. In one of the compound quantities added here, a is to be diminished by b, and in the other, c is to be diminished by d; the *sum* of a and c must therefore be diminished, both by b, and by d, that is, the expression for the the sum total, must contain $-b$ and $-d$. On the same principle, all the quantities which, in the parts to be added, have the negative sign, must

retain this sign, in the amount. Thus $a+2b-c$, added to $d-h-m$, is $a+2b-c+d-h-m$.

68. The sign must be retained also, when a positive quantity is to be added, to a *single* negative quantity. If a be added to $-b$, the sum will be $-b+a$. Here it may be objected, that the negative sign prefixed to b, shows that it is to be *subtracted*. What propriety then can there be in *adding* it? In reply to this, it may be observed, that the sign prefixed to b while standing alone, signifies that b is to be subtracted, *not from a*, but from some *other* quantity, which is not here expressed. Thus $-b$ may represent the *loss*, which is to be subtracted from the *stock* in trade. (Art. 55.) The object of the calculation, however, may not require that the value of this stock should be specified. But the loss is to be connected with a *profit* on some other article. Suppose the profit is 2000 dollars, and the loss 400. The inquiry then is, what is the value of 2000 dollars profit, when connected with 400 dollars loss?

The answer is, evidently, $2000-400$, which shows that 2000 dollars are to be *added* to the stock, and 400 *subtracted* from it; or, which will amount to the same, that the *difference* between 2000 and 400 is to be added to the stock.

69. *Quantities are added, then, by writing them one after another, without altering their signs;* observing always that a quantity, to which *no* sign is prefixed, is to be considered positive. (Art. 29.)

The sum of $a+m$, and $b-8$, and $2h-3m+d$, and $h-n$, and $r+3m-y$, is

$$a+m+b-8+2h-3m+d+h-n+r+3m-y.$$

70. It is immaterial in what *order* the terms are arranged. The sum of a and b and c is either $a+b+c$, or $a+c+b$, or $c+b+a$. For it evidently makes no difference, which of the quantities is added *first*. The sum of 6 and 3 and 9, is the same as 3 and 9 and 6, or 9 and 6 and 3.

And $a+m-n$, is the same as $a-n+m$. For it is plainly of no consequence, whether we first add m to a, and afterwards subtract n; or first subtract n, and then add m.

71. Though connecting quantities by their signs, is all which is *essential* to addition; yet it is desirable to make the expression as simple as may be, by *reducing several terms to one*. The amount of $3a$, and $6b$, and $4a$, and $5b$, is

$$3a+6b+4a+5b$$

But this may be abridged. The first and third terms may be brought into one; and so may the second and fourth.——

For 3 times a, and 4 times a, make 7 times a. And 6 times b, and 5 times b, make 11 times b. The sum, when reduced, is therefore

$$7a+11b.$$

For making the reductions connected with addition, two rules are given, adapted to the two cases, in one of which, the quantities and signs are alike, and in the other, the quantities are alike, but the signs are unlike. Like quantities are the same *powers* of the same *letters*. (Art. 45.) But as the addition of powers and radical quantities will be considered, in a future section, the examples given in this place, will be all of the first power.

72. CASE I. *To reduce several terms to one, when the quantities are alike and the signs alike, add the co-efficients, annex the common letter or letters, and prefix the common sign.*

Thus, to reduce $3b+7b$, that is, $+3b+7b$ to one term, add the co-efficients 3 and 7, to the sum 10, annex the common letter b, and prefix the sign $+$. The expression will then be $+10b$. That 3 times any quantity, and 7 times the same quantity, make 10 times that quantity, needs no proof. In the same manner

$bc+2bc+9bc+3bc$ becomes, when reduced, $15bc$.

And $3xy+7xy+xy+2xy=13xy$. See the two first of the following examples.

bc	$3xy$	$7b+\ xy$	$ry+3abh$	$cdxy+3mg$
$2bc$	$7xy$	$8b+3xy$	$3ry+\ abh$	$2cdxy+\ mg$
$9bc$	xy	$2b+2xy$	$6ry+4abh$	$5cdxy+7mg$
$3bc$	$2xy$	$6b+5xy$	$2ry+\ abh$	$7cdxy+8mg$
$15bc$		$23b+11xy$		$15cdxy+19mg$

The mode of proceeding will be the same, if the signs are *negative*.

Thus $-3bc-bc-5bc$, becomes, when reduced, $-9bc$. And $-ax-3ax-2ax=-6ax$. Or thus,

$-3bc$	$-\ ax$	$-2ab-\ my$	$-3ach-8bdy$
$-\ bc$	$-3ax$	$-\ ab-3my$	$-\ ach-\ bdy$
$-5bc$	$-2ax$	$-7ab-8my$	$-5ach-7bdy$
$-9bc$		$-10ab-12my$	

73. It may perhaps be asked here, as in art. 68, what propriety there is, in *adding* quantities, to which the negative sign is prefixed; a sign which denotes *subtraction?* The answer to this is, that when the negative sign is applied to several quantities, it is intended to indicate that these quantities are to be subtracted, *not from each other,* but from some *other* quantity, marked with the contrary sign. Suppose that, in estimating a man's property, the sum of money in his possession is marked +, and the debts which he owes are marked —. If these debts are 200, 300, 500 and 700 dollars, and if a is put for 100; they will together be $-2a-3a-5a-7a$. And the several terms reduced to one, will evidently be $-17a$, that is, 1700 dollars.

74. CASE II. *To reduce several terms to one, when the quantities are alike, but the signs unlike, take the less co-efficient from the greater; to the difference, annex the common letter or letters, and prefix the sign of the greater co-efficient.*

Thus instead of $8a-6a$, we may write $2a$.

And instead of $7b-2b$, we may put $5b$.

For the simple expression, in each of these instances, is equivalent to the compound one, for which it is substituted.

To	$+6b$	$+4b$	$5bc$	$2hm$	$-dy+6m$	$3h-dx$
Add	$-4b$	$-6b$	$-7bc$	$-9hm$	$4dy-m$	$5h+4dx$
Sum	$+2b$		$-2bc$		$3dy+5m$	

75. Here again, it may excite surprise, that what appears to be subtraction, should be introduced under addition. But according to what has been observed, (Art. 66,) this subtraction is, strictly speaking, no part of the addition. It belongs to a consequent *reduction.* Suppose $6b$ is to be added to $a-4b$. The sum is

$$a-4b+6b.$$

But this expression may be rendered more simple. As it now stands, $4b$ is to be subtracted from a, and $6b$ added.— But the amount will be the same, if, without subtracting any thing, we add $2b$, making the whole $a+2b$. And in all similar instances, the *balance* of two or more quantities, may be substituted for the quantities themselves.

76. The co-efficient of a sum when reduced to one term, may be less, than either of the co-efficients of the quantities which are thus reduced. In one of the preceding ex-

amples, $6b-4b=2b$. Here 2, the co-efficient of the single term, is less than 6 or 4, the co-efficients of the two terms, to which the single one is equal. The balance of a book account may be less, than either the debt, or the credit. It may even be nothing. Hence,

77. If two *equal* quantities have *contrary signs*, they destroy each other, and may be cancelled. Thus $+6b-6b$ $=0$: And $\overline{3\times6}-18=0$: And $7bc-7bc=0$.

Let there be any two quantities whatever, of which a is the greater, and b the less.

$$\text{Their sum will be} \quad a+b$$
$$\text{And their difference} \quad a-b$$

The sum and difference added, will be $2a+0$, or simply $2a$. That is, if the *sum* and *difference* of any two quantities be added together, the *whole* will be *twice* the greater quantity. This is one instance, among multitudes, of the rapidity with which *general* truths are discovered and demonstrated in algebra. (Art. 23.)

78. If several positive, and several negative quantities are to be reduced to one term; first reduce those which are positive, next those which are negative, and then take the *difference* of the co-efficients, of the two terms thus found.

Ex. 1. Reduce $13b+6b+b-4b-5b-7b$, to one term.

$$\text{By art. 72, } 13b+6b+ b= 20b \Big\}$$
$$\text{And} \quad -4b-5b-7b=-16b$$

By art. 74 $\quad 20b-16b= \quad 4b$, which is the value of all the given quantities, taken together.

Ex. 2. Reduce $3xy-xy+2xy-7xy+4xy-9xy+7xy-6xy$.

The positive terms are		The negative terms are
$3xy$		$-xy$
$2xy$		$-7xy$
$4xy$		$-9xy$
$7xy$		$-6xy$
And their sum is $16xy$		$-23xy$

Then $16xy-23xy=-7xy$.

Ex. 3. $3ad-6ad+ad+7ad-2ad+9ad-8ad-4ad=0.$

4. $2abm-abm+7abm-3abm+7abm=$

5. $axy-7axy+8axy-axy-8axy+9axy=$

79. If the *letters*, in the several terms to be added, are different, they can only be placed after each other, with their proper signs. They cannot be united in one simple term. If $4b$, and $-6y$, and $3x$, and $17h$, and $-5d$, and 6, be added; their sum will be

$$4b-6y+3x+17h-5d+6. \quad (\text{Art. } 69.)$$

Different letters can no more be united in the same term, than dollars and guineas can be added, so as to make a single sum. Six guineas and 4 dollars are neither ten guineas nor ten dollars. Seven hundred, and five dozen are neither 12 hundred nor 12 dozen. But, in such cases, the algebraic signs serve to show how the different quantities stand related to each other; and to indicate future operations, which are to be performed, whenever the letters are converted into numbers. In the expression $a+6$, the two terms can not be united in one. But if a stands for 15, and if, in the course of a calculation, this number is restored; then $a+6$ will become $15+6$, which is equivalent to the single term 21. In the same manner, $a-6$ becomes $15-6$, which is equal to 9. The signs keep in view the relations of the quantities, till an opportunity occurs of reducing several terms to one.

80. When the quantities to be added contain several terms which are *alike*, and several which are *unlike*, it will be convenient to arrange them in such a manner, that the similar terms may stand one under the other.

To $\quad 3bc-6d+2b-3y$) These may be arranged thus.
Add $-3bc+x-3d+bg$ } $\quad 3bc-6d+2b-3y$
And $\quad 2d+y+3x+b$) $-3bc-3d \qquad + x+bg$
$$\qquad\qquad\qquad 2d \quad + y+3x \quad +b$$

The sum will be $\qquad -7d+2b-2y+4x+bg+b.$

In the first term, $3bc$ is balanced by $-3bc$, so that this term disappears in the general amount. (Art. 77.)

F

EXAMPLES.

1. Add and reduce $ab+8$ to $cd-3$ and $5ab-4m+2$.
The sum is $6ab+7+cd-4m$.

2. Add $x+3y-dx$, to $7-x-8+hm$.
Ans. $3y-dx-1+hm$.

3. Add $abm-3x+bm$, to $y-x+7$, and $5x-6y+9$.
Ans.

4. Add $3am+6-7xy-8$, to $10xy-9+5am$.
Ans.

5. Add $6ahy+7d-1+mxy$, to $3ahy-7d+17-mxy$.
Ans.

6. Add $7ad-h+8xy-ad$, to $5ad+h-7xy$.
Ans.

SUBTRACTION.

Art. 81. ADDITION is bringing quantities together, to find their amount. On the contrary, SUB-TRACTION *is finding the* DIFFERENCE *of two quantities, or sets of quantities.*

Particular rules might be given, for the several cases in subtraction. But it is more convenient to have one general rule, founded on the principle, that *taking away* a *positive* quantity, from an algebraic expression, is the same in effect, as *annexing* an equal *negative* quantity; and taking away a negative quantity is the same, as annexing an equal positive one.

Suppose $+b$ is to be subtracted from $a+b.$
Taking away $+b$, from $a+b$, leaves $a.$
And annexing $-b$, to $a+b$, gives $a+b-b.$
But by axiom 5th, $a+b-b$ is equal to $a.$

That is, *taking away* a *positive* term, from an algebraic expression, is the same in effect, as *annexing* an equal *negative* term.

Again, suppose $-b$ is to be subtracted from $a-b$
Taking away $-b$, from $a-b$, leaves a
And annexing $+b$, to $a-b$, gives $a-b+b$
But $a-b+b$ is equal to a

That is, *taking away* a *negative* term, is equivalent to *annexing* a *positive* one. If an estate is encumbered with a debt; to cancel this debt, is to add so much to the value of the estate. Subtracting an item from one side of a book-account, will produce the same alteration in the balance, as adding an equal sum to the opposite side.

To place this in another point of view.
If m is *added* to b, the sum is by the notation, $b+m$ ⎫
But if m is subtracted from b, the remainder is $b-m$ ⎬
So if m and h are each added to b, the sum is $b+m+h$ ⎫
But if m and h are each subtracted from b, the rem'dr is $b-m-h$ ⎬

The only difference then between adding a positive quantity and subtracting it, is, that the *sign* is changed from + to —.

Again, if $m-n$ is subtracted from b, the remainder is,

$$b-m+n$$

For the *less* the quantity subtracted, the *greater* will be the remainder. But in the expression $m-n$, m is diminished by n; therefore, $b-m$ must be *increased* by n; so as to become $b-m+n$: that is, $m-n$ is subtracted from b, by changing $+m$ into $-m$, and $-n$ into $+n$, and then writing them after b, as in addition. The explanation will be the same, if there are *several* quantities which have the negative sign. Hence,

82. *To perform subtraction in algebra, change the signs of all the quantities to be subtracted, or suppose them to be changed, from + to —, or from — to +, and then proceed as in addition.*

The signs are to be changed, in the *subtrahend* only.— Those in the minuend are not to be altered. Although the rule here given is adapted to every case of subtraction; yet there may be an advantage in giving some of the examples in distinct classes.

83. In the first place, the signs may be *alike*, and the minuend *greater* than the subtrahend.

From	$+28$	$16b$	$14da$	-28	$-16b$	$-14da$
Subtract	$+16$	$12b$	$6da$	-16	$-12b$	$-6da$
Difference	$+12$	$4b$	$8da$	-12	$-4b$	$-8da$

Here, in the first example, the + before 16 is supposed to be changed into —, and then, the signs being unlike, the two terms are brought into one, by the second case of reduction in addition. (Art. 74.) The two next examples are subtracted in the same way. In the three last, the — in the subtrahend, is supposed to be changed into +. It may be well for the learner, at first, to write out the examples; and actually to change the signs, instead of merely conceiving them to be changed. When he has become familiar with the operation, he can save himself the trouble of transcribing.

84. In the second place, the signs may be alike, and the minuend *less* than the subtrahend.

From	$+16b$	$12b$	$6da$	-16	$-12b$	$-6da$
Sub.	$+28b$	$16b$	$14da$	-28	$-16b$	$-14da$
Dif.	-12	$-4b$	$-8da$	$+12$	$4b$	$8da$

The same quantities are given here, as in the preceding article, for the purpose of comparing them together. But the minuend and subtrahend are made to change places. The mode of subtracting is the same. In this class a *greater* quantity is taken from a *less*: in the preceding, a *less* from a *greater*. By comparing them, it will be seen, that there is no difference in the answers, except that the *signs* are *opposite*. Thus $16b-12b$ is the same as $12b-16b$, except that one is $+4b$, and the other $-4b$: That is, a greater quantity subtracted from a less, gives the same result, as a less subtracted from a greater, except that the one is positive and the other negative. See art. 58 and 59.

85. In the third place, the *signs* may be *unlike*.

From	$+28$	$+16b$	$+14da$	-28	$-16b$	$-14da$
Sub.	$+16$	$-12b$	$-6da$	$+16$	$+12b$	$+6da$
Dif.	$+44$	$28b$	$20da$	-44	-28	$-20da$

From these examples, it will be seen that the *difference* between a positive and a negative quantity, may be *greater* than either of the two quantities. In the first example, 44 the difference is greater, than 28 the minuend, or 16 the subtrahend. In a thermometer, the difference between 28 degrees above cypher, and 16 below, is 44 degrees. The difference between gaining 1000 dollars in trade and losing 500, is equivalent to 1500 dollars.

86. Subtraction may be *proved*, as in arithmetic, by adding the remainder to the subtrahend. The sum ought to be equal to the minuend, upon the obvious principle, that the difference of two quantities added to one of them, is equal to the other. This serves not only to correct any particular errour, but to verify the general rule.

From	$3abm-\ xy$	$-17+4ax$	$ax+\ 7b$	$3ah+axy$
Sub.	$-7abm+6xy$	$-20-\ ax$	$-4ax+15b$	$-7ah+axy$

| Rem. | $10abm-7xy$ | | $5ax-\ 8b$ | |

87. When there are *several terms alike*, they may be reduced as in addition.

1. From ab, subtract $3am+am+7am+2am+6am$.
Ans. $ab-3am-am-7am-2am-6am=ab-19am$. (Art.72.)

2. From y, subtract $-a-a-a-a$.
Ans. $y+a+a+a+a=y+4a$.

3. From $ax-bc+3ax+7bc$, subtract $4bc-2ax+bc+4ax$.
Ans. $ax-bc+3ax+7bc-4bc+2ax-bc-4ax=2ax+bc$.
(Art. 78.)

4. From $ad+3dc-bx$, subtract $3ad+7bx-dc+ad$.
Ans.

88. When the *letters* in the minuend are different from those in the subtrahend, the latter are subtracted, by first changing the signs, and then placing the several terms one after another, as in addition. (Art. 79.)

1. From $3ab+8-my+dh$, subtract $x-dr+4hy-bmx$.
Ans. $3ab+8-my+dh-x+dr-4hy+bmx$.

2. $13ad+xy+d-(7ad-xy+d+hm-ry)=6ad+2xy-hm+ry$.

3. $7abc-8+7x-(3abc-8-dx+r)=4abc+7x+dx-r$.

4. $3ad+h-2y-(7y+3h-mx+4ad-hy-ad)=$

5. $6am-dy+8-(16+3dy-8+am-d+r)=$

MULTIPLICATION.*

ART. 89. $\mathbf{I}$N addition, one quantity is connected with another. It is frequently the case, that the quantities brought together are *equal;* that is, a quantity is added to *itself.*

As $3+3=6$ $3+3+3+3=12$

     $3+3+3=9$ $3+3+3+3+3=15$, &c.

And $a+a=2a$ $a+a+a+a=4a$

     $a+a+a=3a$ $a+a+a+a+a=5a$, &c.

This repeated addition of a quantity to itself, is what was, originally, called *multiplication.* But the term, as it is now used, has a more extensive signification. We have frequent occasion to repeat, not only the *whole* of a quantity, but a certain *portion* of it. If the stock of an incorporated company is divided into shares, one man may own ten of them, another five, and another a *part* only of a share, say two fifths. When a dividend is made, of a certain sum on a share, the first is entitled to *ten* times this sum, the second to *five* times, and the third to only *two fifths* of it. As the apportioning of the dividend, in each of these instances, is upon the same principle, it is called multiplication in the last, as well as in the two first.

Again, suppose a man is obligated to pay an annuity of 100 dollars a year. As this is to be *subtracted* from his estate, it may be represented by $-a$. And as it is to be subtracted *year after year.* it will become, in four years, $-a-a-a-a$ $=-4a$. This *repeated subtraction* is also called multiplication. According to this view of the subject;

* Newton's Universal Arithmetic, p. 4. Maseres on the Negative Sign, Sec. II. Camus' Arithmetic, Book II. Chap. 3. Euler's Algebra, Sec. I. and II. Chap. 3. Simpson's Algebra, Sec. IV. Maclaurin, Saunderson, Lacroix, Ludlam.

90. *Multiplying by a whole number is taking the multipli-cand as many times, as there are units in the multiplier.*

Multiplying by 1, is taking the multiplicand *once*, as a.

Multiplying by 2, is taking the multiplicand *twice*, as $a+a$.

Multiplying by 3, is taking the mult'd *three times,*as $a+a+a$,&c.

Multiplying by a FRACTION *is taking a certain* PORTION *of the multiplicand as many times, as there are like portions of an unit in the multiplier.*

Multiplying by $\frac{1}{3}$, is taking $\frac{1}{3}$ of the mult'd *once*, as $\frac{1}{3}a$.

Multiplying by $\frac{2}{3}$, is taking $\frac{1}{3}$ of the mult'd *twice*, as $\frac{1}{3}a+\frac{1}{3}a$.

Multipl'g by $\frac{3}{3}$, is taking $\frac{1}{3}$ of the mult'd *three times,*as $\frac{1}{3}a+\frac{1}{3}a+\frac{1}{3}a$.

Hence, if the multiplier is *an unit*, the product is *equal* to the multiplicand: If the multiplier is *greater* than an unit, the product is greater than the multiplicand: And if the multiplier is *less* than an unit, the product is less than the multiplicand.

Multiplication by a NEGATIVE *quantity, has the same relation to multiplication by a positive quantity, which* SUBTRACTION *has to addition.* In the one, the sum of the repetitions of the multiplicand is to be *added*, to the other quantities with which the multiplier is connected. In the other, the sum of these repetitions is to be *subtracted* from the other quantities. This subtraction is performed at the time of multiplying, by changing the sign of the product. See Art. 107 and 108.

91. Every multiplier is to be considered a *number*. We sometimes speak of multiplying by a given *weight*, or *measure*, a sum of *money*, &c. But this is abbreviated language. If construed literally, it is absurd. Multiplying is taking either the whole or a part of a quantity, a certain *number of times*. To say that one quantity is repeated as many times, as another is *heavy*, is nonsense. But if a part of the weight of a body be fixed upon as *an unit*, a quantity may be multiplied by a *number* equal to the number of these parts contained in the body. If a diamond is sold by weight, a particular price may be agreed upon for each *grain*. A grain is here *the unit;* and it is evident that the value of the diamond, is equal to the given price repeated as many times, as there are grains in the whole weight. We say concisely that the price is multiplied by the *weight;* meaning that it is multiplied by a *number* equal to the number of grains in the weight. In a similar manner, any quantity whatever

---

* See note A. at the end.

may be supposed to be made up of parts, each being con-sidered *a unit*, and any number of these may become a multiplier.

 92. As multiplying is taking the whole or a part of a quantity a certain number of times, it is evident that the *product* must be of the same nature as the *multiplicand.*

If the multiplicand is an abstract *number;* the product will be a number.

If the multiplicand is *weight,* the product will be weight. If the multiplicand is a *line,* the product will be a line. *Repeating* a quantity does not alter its nature. It is frequently said, that the product of two lines is a *surface,* and that the product of three lines is a *solid.* But these are abridged expressions, which if interpreted literally are not correct. See the section on the application of Algebra to Geometry.

93. The multiplication of *fractions* will be the subject of a future section. We have first to attend to multiplication by positive whole numbers. This, according to the definition (Art. 90) is taking the multiplicand as many times, as there are units in the multiplier. Suppose a is to be multiplied by b, and that b stands for 3. There are, then, three units in the multiplier b. The multiplicand must therefore be taken three times; thus,

$$a$$
$$a$$
$$a$$
$$\overline{}$$

The amount is $3a$, that is, 3 times a; which, if b stands for 3, is the same as $b \times a$ or ba. (Art. 40.) Or thus, $a+a+a =3a$, which is the same as ba. So that, *multiplying two letters together is nothing more, than writing them one after the other,* either with, or without, the sign of multiplication between them. Thus b multiplied into c, is $b \times c$, or bc. And x into y, is $x \times y$, or $x.y$, or xy.

94. If more than two letters are to be multiplied, they must be connected in the same manner. Thus a into b and c, is cba. For by the last article, a into b, is ba. This product is now to be multiplied into c. If c stands for 5, then ba is to be taken five times, thus,

$$ba+ba+ba+ba+ba=5ba, \text{ or } cba.$$

The same explanation may be applied to any number of letters. Thus am into xy, is $amxy$. And bh into mrx, is $bhmrx$.

95. It is immaterial *in what order* the letters are arranged.

G

The product ba is the same as ab. Three times five is equal to five times three. Let the number 5 be represented by as many points, in a *horizontal* line; and the number 3, by as many points in a *perpendicular* line.

Here it is evident that the *whole* number of points is equal, either to the number in the *horizontal* row *three* times repeated, or to the number in the *perpendicular* row *five* times repeated; that is to 5×3, or 3×5. This explanation may be extended to a series of factors consisting of any numbers whatever. For the product of two of the factors may be considered as one number. This may be placed before or after a third factor: the product of three, before or after a fourth, &c.

Thus $24 = 4 \times 6$ or $6 \times 4 = 4 \times 3 \times 2$ or $4 \times 2 \times 3$ or $2 \times 3 \times 4$.
The product of a, b, c, and d, is $abcd$, or $acdb$, or $dcba$, or $badc$.
It will generally be convenient, however, to place the letters in *alphabetical* order.

96. *When the letters have numerical* co-efficients, *these must be multiplied together, and prefixed to the product of the letters.*

Thus $3a$ into $2b$ is $6ab$. For if a into b is ab, then 3 times a into b, is evidently $3ab$: and if, instead of multiplying by b, we multiply by *twice* b, the product must be twice as great, that is $2 \times 3ab$ or $6ab$.

Mult'y	$9ab$	$12hy$	$3dh$	$2ad$	$7bdh$	$3ay$
Into	$3xy$	$2rx$	my	$13hmg$	x	$8mx$
Prod.	$27abxy$		$3dhmy$		$7bdhx$	

97. If either of the factors consists of figures *only*, these must be multiplied into the co-efficients and letters of the other factors.

Thus $3ab$ into 4, is $12ab$. And 36 into $2x$, is $72x$. And 24 into hy, is $24hy$.

98. If the multiplicand is a *compound* quantity, *each of its terms must be multiplied into the multiplier*. Thus $b+c+d$ into a is $ab+ac+ad$. For the whole of the multiplicand is to be taken as many times, as there are units in the multi-

plier. If then a stands for 3, the repetitions of the multiplicand are

$$b+c+d$$
$$b+c+d$$
$$b+c+d$$

And their sum is $3b+3c+3d$, that is, $ab+ac+ad$.

Mult.	$d+2xy$	$2h+m$	$3hl+1$	$2hm+3+dr$
Into	$3b$	$7dy$	my	$4b$
Prod.	$3bd+6bxy$		$3hlmy+my$	

99. The preceding instances must not be confounded
with those in which several *factors* are connected by the
sign $\times$, or by a point. In the latter case, the multiplier is
to be written before the other factors *without being repeated*.
The product of $b \times d$ into a, is $ab \times d$, and not $ab \times ad$. For
$b \times d$ is bd, and this into a, is abd. (Art. 94.) The expression
$b \times d$ is not to be considered, like $b+d$, a *compound quantity*
consisting of two terms. Different terms are always separated by $+$ or $-$. (Art. 36.) The product of $b \times h \times m \times y$
into a, is $a \times b \times h \times m \times y$ or $abhmy$. But $b+h+m+y$ into a,
is $ab+ah+am+ay$.

100. If *both* the factors are compound quantities, *each*
term in the multiplier must be multiplied into each in the multiplicand.

Thus $\overline{a+b}$ into $\overline{c+d}$ is $ac+ad+bc+bd$.

For the units in the multiplier $a+b$ are equal to the units
in a added to the units in b. Therefore the product produ-
ced by a, must be added to the product produced by b.
The product of $c+d$ into a is $ac+ad$ $\Big\}$ Art. 98.
The product of $c+d$ into b is $bc+bd$
The product of $c+d$ into $a+b$ is therefore $ac+ad+bc+bd$.

Mult.	$3x+d$		$4ay+2b$	$a+1$
Into	$2a+hm$		$3c+rx$	$3x+4$
Prod.	$6ax+2ad+3hmx+dhm$			$3ax+3x+4a+4$

Mult. $2h+7$ into $6d+1$.　Prod. $12dh+42d+2h+7$.

Mult. $dy+rx+h$ into $6m+4+7y$.　Prod.

Mult. $7+6b+ad$ into $3r+4+2h$.　Prod.

　101. When several terms in the product are *alike*, it will be expedient to *set one under the other*, and then to unite them, by the rules for reduction in addition.

Mult. $b+a$　　　$b+c+2$　　　$a+\ y+1$
Into　$b+a$　　　$b+c+3$　　　$3b+2x+7$

$bb+ab$　　　$bb+bc+2b$
　$+ab+aa$　　　$bc\quad +cc+2c$
　　　　　　　　$+3b\quad +3c+6$

Prod. $bb+2ab+aa$　$bb+2bc+5b+cc+5c+6$

Mult. $3a+d+4$ into $2a+3d+1$.　Prod.

Mult. $b+cd+2$ into $3b+4cd+7$.　Prod.

　102. Here, as in Art. 99, care must be taken not to confound *terms* with *factors*.

The product of $a\times b$ into $c\times d$, is $a\times b\times c\times d$, or $abcd$.

But the product of $a+h$ into $c+d$, is $ac+ad+bc+bd$.

The product of $3b+2c$ into $h\times m$, is $3bhm+2chm$.

The product of $a\times b\times c$ into $h+4y$, is $abch+4abcy$.

　103. It will be easy to see that when the multiplier and multiplicand consist of any quantity *repeated as a factor*, this factor will be repeated in the product, as many times as in the multiplier and multiplicand together.

Mult. $a\times a\times a$　Here a is repeated *three times* as a factor.
Into　$a\times a$　　Here it is repeated *twice*.

Prod. $a\times a\times a\times a\times a$ Here it is repeated *five times*.

　The product of $bbbb$ into bbb, is $bbbbbbb$.

　The product of $2x\times 3x\times 4x$ into $5x\times 6x$, is $2x\times 3x\times 4x\times 5x\times 6x$.

　104. But the *numeral co-efficients* of several fellow-factors may be brought together by multiplication.

　Thus $2a\times 3b$ into $4a\times 5b$ is $2a\times 3b\times 4a\times 5b$, or $120aabb$,

For the co-efficients are *factors,* (Art. 41,) and it is immaterial in what *order* these are arranged. (Art. 95.) So that
$2a \times 3b \times 4a \times 5b = 2 \times 3 \times 4 \times 5 \times a \times a \times b \times b = 120aabb.$

The product of $3a \times 4bh$ into $5m \times 6y$, is $360\ abhmy$.

The product of $4b \times 6d$ into $2x+1$, is $48bdx + 24bd$.

105. The examples in multiplication thus far have been confined to *positive* quantities. It will now be necessary to consider, in what manner the result will be affected, by multiplying positive and negative quantities together. We shall find,

That $+$ into $+$ produces $+$

$-$ into $+$ $-$

$+$ into $-$ $-$

$-$ into $-$ $+$

All these may be comprised in one general rule, which it will be important to have always familiar. *If the signs of the factors are* ALIKE, *the sign of the product will be affirmative; but if the signs of the factors are* UNLIKE, *the sign of the product will be negative.*

106. The first case, that of $+$ into $+$, needs no farther illustration. The second is $-$ into $+$, that is, the multiplicand is negative, and the multiplier positive. Here $-a$ into $+4$ is $-4a$. For the repetitions of the multiplicand are,
$$-a-a-a-a = -4a.$$

Mult.	$b-3a$	$2a-m$	$h-3d-4$	$a-2-7d-x$
Into	$6y$	$3h+x$	$2y$	$3b+h$
Prod.	$6by-18ay$		$2hy-6dy-8y$	

107. In the two preceding cases, the affirmative sign prefixed to the multiplier shows, that the repetitions of the multiplicand are to be *added,* to the other quantities with which the multiplier is connected. But in the two remaining cases, the negative sign prefixed to the multiplier, indicates that the sum of the repetitions of the multiplicand are to be *subtracted* from the other quantities. (Art. 90.) And this subtraction is performed, at the time of multiplying, by making the sign of the product opposite to that of the multiplicand. Thus $+a$ into -4, is $-4a$. For the repetitions of the multiplicand are,
$$+a+a+a+a = +4a.$$
But this sum is to be *subtracted,* from the other quantities

with which the multiplier is connected. It will then become
—4a. (Art. 82.)

Thus, in the expression $b-(4 \times a)$, it is manifest that $4 \times a$
is to be subtracted from b. Now $4 \times a$ is $4a$, that is, $+4a$.
But, to subtract this from b, the sign $+$ must be changed into $-$. So that $b-(4 \times a)$ is $b-4a$. And $a \times -4$ is therefore —4a.

Again, suppose the multiplicand is a, and the multiplier
$(6-4)$. As $(6-4)$ is equal to 2, the product will be equal to
$2a$. This is *less* than the product of 6 into a. To obtain
then the product of the compound multiplier $(6-4)$ into a,
we must *subtract* the product of the negative part, from
that of the positive part.

$$\left. \begin{array}{l} \text{Multiplying} \quad a \\ \text{Into} \quad\quad 6-4 \end{array} \right\} \text{ is the same as } \left\{ \begin{array}{l} \text{Multiplying } a \\ \text{Into} \quad\quad 2 \end{array} \right.$$

And the prod. $6a-4a$, is the same as the product $2a$.
Therefore a into -4, is $-4a$.

But if the multiplier had been $(6+4)$, the two products
must have been *added*.

$$\left. \begin{array}{l} \text{Multiplying} \quad a \\ \text{Into} \quad\quad 6+4 \end{array} \right\} \text{ is the same as } \left\{ \begin{array}{l} \text{Multiplying } a \\ \text{Into} \quad\quad 10 \end{array} \right.$$

And the prod. $6a+4a$ is the same as the product $10a$

This shows at once the difference between multiplying by
a *positive* factor, and multiplying by a *negative* one. In the
former case, the sum of the repetitions of the multiplicand
is to be *added to*, in the latter, *subtracted from*, the other
quantities, with which the multiplier is connected. For every negative quantity must be supposed to have a reference
to some other which is positive; though the two may not always stand in connection, when the multiplication is to be
performed.

Mult.	$a+b$		$3dy+hx+2$		$3h\ +3$
Into	$b-x$		$mr-ab$		$ad-6$

Prod. $ab+bb-ax-bx$ $\qquad\qquad 3adh+3ad-18h-18$

108. If *two negatives* be multiplied together, the product

will be affirmative :$-4\times-a=+4a$. In this case, as in the preceding, the repetitions of the multiplicand are to be *subtracted*, because the multiplier has the negative sign. These repetitions, if the multiplicand is $-a$ and the multiplier -4, are $-a-a-a-a=-4a$. But this is to be subtracted by changing the sign. It then becomes $+4a$.

Suppose $-a$ is multiplied into $(6-4)$. As $6-4=2$, the product is evidently, *twice* the multiplicand, that is $-2a$. But if we multiply $-a$ into 6 and 4 separately; $-a$ into 6 is $-6a$, and $-a$ into 4 is $-4a$. (Art. 106.) As, in the multiplier, 4 is to be subtracted from 6; so, in the product, $-4a$ must be subtracted from $-6a$. Now $-4a$ becomes by subtraction $+4a$. The whole product then is $-6a+4a$, which is equal to $2a$. Or thus,

$$\text{Multiplying} \quad -a \atop \text{Into} \qquad 6-4 \Big\} \text{ is the same as } \Big\{ \text{Multiplying} \ -a \atop \text{Into} \qquad\quad 2$$

And the prod. $-6a+4a$, is equal to the product $-2a$.

In Double Fellowship in arithmetic, each man's stock is to be multiplied into the time for which it is employed. Suppose there are two partners A and B; that B's stock is 300 dollars less than A's; and that the time of the former is two years less, than that of the latter : then

If A's share is equal to c; B's share will equal $c-300$?
And If A's time is equal to d; B's time will equal $d-2$

$$\text{Multiplying} \ c-300 \atop \text{Into} \qquad d-2$$

The product will be $cd-300d-2c+600$.

Here the two first terms are obtained, by multiplying $(c-300)$ into d; that is, B's stock into the *whole* time represented by d. But this time is two years too much. The product is therefore *too great*. It ought to be diminished, by the product of the stock $(c-300)$ into 2. The whole product will then be

$(cd-300d)-(2c-600)=cd-300d-2c+600$.

Here $2c-600$ is subtracted by changing the signs (Art. 88.) so that -300×-2 is $+600$. On the same principle, it is necessary to change the signs of every term in a compound quantity which is multiplied by a negative factor.

It is often considered a great mystery, that the product of two negatives should be affirmative. But it amounts to nothing more than this, that the subtraction of a negative quantity, is equivalent to the addition of an affirmative, (Art. 81,) and, therefore, that the *repeated* subtraction of a negative, is equivalent to a *repeated* addition of an affirmative. Taking off from a man's hands a debt of ten dollars every month, is adding ten dollars a month to the value of his property.

| Mult. | $a-4$ | | $3d-hy-2x$ | $3ay-b$ |
| Into | $3b-6$ | | $4b-7$ | $6x-1$ |

Prod. $3ab-12b-6a+24$ $18axy-6bx-3ay+b$

Multiply $3ad-ah-7$ into $4-dy-hr$.

Multiply $2hy+3m-1$ into $4d-2x+3$.

109. As a negative multiplier changes the sign of the quantity which it multiplies; if there are *several* negative factors to be multiplied together,

The *two first* will make the product *positive*;

The *third* will make it *negative*;

The *fourth* will make it *positive*, &c.

Thus $-a$ And $-abc$

Into $-b$ Into $-d$

——   [factors.

Gives $+ab$, the prod. of *two*   Gives $+abcd$, *four* fact's.

This into $-c$ This into $-e$

Gives $-abc$, *three* factors.   Gives $-abcde$, *five* fact's.

That is,

The product of any *even* number of negative factors is *positive;* but the product of any *odd* number of negative factors is *negative*.

Thus $-a\times-a=aa$ And $-a\times-a\times-a\times-a=aaaa$

$-a\times-a\times-a=-aaa$ $-a\times-a\times-a\times-a\times-a=-aaaaa$.

The product of several factors which are *all positive*, is invariably positive.

110. Positive and negative terms may frequently *balance*

each other, so as to disappear in the product. (Art. 77.) A star is sometimes put in the place of the deficient term.

Mult. $a-b$	$mm-yy$	$aa+ab+bb$
Into $a+b$	$mm+yy$	$a-b$

$$
\begin{array}{ccc}
aa-ab & & aaa+aab+abb \\
+ab-bb & & -aab-abb-bbb
\end{array}
$$

Prod. $aa \quad * \quad -bb$ $\qquad$ $aaa \quad * \quad * \quad -bbb$

111. For many purposes, it is sufficient merely to *indicate* the multiplication of compound quantities, without actually multiplying the several terms. Thus the product of
$a+b+c$ into $h+m+y$, is $(a+b+c)\times(h+m+y)$. (Art. 40.)
The product of
$a+m$ into $h+x$ and $d+y$, is $(a+m)\times(h+x)\times(d+y)$.
By this method of representing multiplication, an important advantage is often gained, in preserving the factors distinct from each other.

When the several terms are multiplied in form, the expression is said to be *expanded*. Thus   [Art. 100.
$(a+b)\times(c+d)$ becomes when expanded $ac+ad+bc+bd$.

112. With a given multiplicand, the less the multiplier, the less will be the product. If then the multiplier be reduced to *nothing*, the *product* will be nothing. Thus $a\times0=0$. And if 0 be one of *any number* of fellow-factors, the product of the whole will be nothing.
Thus $ab\times c\times 3d\times 0=3abcd\times 0=0$.
And $(a+b)\times(c+d)\times(h-m)\times 0=0$.

113. Although, for the sake of illustrating the different points in multiplication, the subject has been drawn out into a considerable number of particulars; yet it will scarcely be necessary for the learner, after he has become familiar with the examples, to burden his memory with any thing more than the following general rule.

Multiply the letters and co-efficients of each term in the multiplicand, into the letters and co-efficients of each term in the multiplier; and prefix, to each term of the product, the sign required by the principle, that like signs produce +, and different signs —.

H

Mult. $a+3b-2$ into $4a-6b-4$.

Mult. $4ab \times x \times 2$ into $3my-1+h$.

Mult. $(7ah-y) \times 4$ into $4x \times 3 \times 5 \times d$.

Mult. $(6ab-hd+1) \times 2$ into $(8+4x-1) \times d$.

Mult. $3ay+y-4+h$ into $(d+x) \times (h+y)$.

DIVISION.

Art. 114. IN multiplication, we have two factors given, and are required to find their product. By multiplying the factors 4 and 6, we obtain the product 24. But it is frequently necessary to reverse this process. The number 24, and *one* of the factors, may be given, to enable us to find the other. The operation by which this is effected is called *Division*. We obtain the number 4, by dividing 24 by 6. The quantity to be divided is called the dividend; the *given* factor, the divisor; and that which is *required*, the quotient.

115. DIVISION *is finding a quotient, which multiplied into the divisor will produce the dividend.**

In multiplication, the *multiplier* is always a *number*. (Art. 91.) And the *product* is a quantity of the same kind, as the multiplicand. (Art. 92.) The product of 3 rods into 4, is 12 rods. When we come to division, the product and *either* of the factors may be given, to find the other: that is,

The divisor may be a *number*, and then the quotient will be a quantity of the same kind as the dividend; or

The *divisor* may be a quantity of the same kind as the dividend; and then the *quotient* will be a number.

Thus $\dfrac{12\ rods}{4} = 3\ rods.$ But $\dfrac{12\ rods}{3\ rods} = 4$

And $\dfrac{12\ rods}{24} = \frac{1}{2}\ rod.$ And $\dfrac{12\ rods}{24\ rods} = \frac{1}{2}$

In the first case, the divisor, being a *number*, shows into *how many* parts the dividend is to be separated; and the quotient shows what these parts are.

---

* The *remainder* is here supposed to be included in the quotient, as is commonly the case in algebra.

If 12 rods be divided into 4 parts, each will be 3 rods long.
And if 12 rods be divided into 24 parts, each will be *half* a rod long.

In the other case, if the divisor is *less* than the dividend, the former shows into *what* parts the latter is to be divided; and the quotient shows *how many* of these parts are contained in the dividend. In other words, division in this case consists in finding *how often one quantity is contained in another*.

A line of 3 rods, is contained in one of 12 rods, *four times.*

But if the divisor is *greater* than the dividend, and yet a quantity of the same kind, the quotient shows *what part* of the divisor is equal to the dividend.

Thus *one half* of 24 rods is equal to 12 rods.

116. As the product of the divisor and quotient is equal to the dividend, the quotient may be found, by resolving the dividend into two such factors, that one of them shall be the divisor. The other will, of course, be the quotient.

Suppose *abd* is to be divided by *a*. The factors *a* and *bd* will produce the dividend. The first of these, being a divisor, may be set aside. The other is the quotient. Hence,

When the divisor is found as a factor, in the dividend, the division is performed, by CANCELLING *this factor.*

Divide	cx	dh	drx	hmy	$dhxy$	$abcd$	$abxy$
By	c	d	dr	hm	dy	b	ax
Quot.	x	j	x		hx		by

In each of these examples, the letters which are common to the divisor and dividend, are set aside; and the other letters form the quotient. It will be seen at once, that the product of the quotient and divisor is equal to the dividend.

117. If a letter is *repeated* in the dividend, care must be taken that the factor rejected be only equal to the divisor.

Div.	aab	bbx	$aadddx$	$aammyy$	$aaaxxxh$	yyy
By	a	b	ad	amy	$aaxx$	yy
Quot.	ab		$addx$		axh	

In such instances, it is obvious that we are not to reject

every letter in the dividend which is the same with one in the divisor.

118. If the dividend consists of *any factors whatever*, expunging one of them is dividing by it.

Div.	$a(b+d)$	$a(b+d)$	$(b+x)(c+d)$	$(b+y)\times(d-h)x$
By	a	$b+d$	$b+x$	$d-h$
Quot.	$b+d$	a	$c+d$	$(b+y)\times x$

In all these instances the product of the quotient and divisor is equal to the dividend by Art. 111.

119. In performing multiplication, if the factors contain *numeral figures*, these are multiplied into each other. (Art. 96.) Thus $3a$ into $7b$ is $21ab$. Now if this process is to be *reversed*, it is evident that dividing the number in the product, by the number in one of the factors, will give the number in the other factor. The quotient of $21ab \div 3a$ is $7b$. Hence,

In division, if there are *numeral co-efficients* prefixed to the letters, *the co-efficient of the dividend must be divided, by the co-efficient of the divisor.*

Div.	$6ab$	$16dxy$	$25dhr$	$12xy$	$34drx$	$20hm$
By	$2b$	$4dx$	dh	6	34	m
Quot.	$3a$		$25r$		drx	

120. When a simple factor is multiplied into a *compound* one, the former enters into *every* term of the latter. (Art. 98.) Thus a into $b+d$, is $ab+ad$. Such a product is easily resolved again into its original factors.

Thus $ab+ad=a\times(b+d)$
$ab+ac+ah=a\times(b+c+h)$
$amh+amx+amy=am\times(h+x+y)$
$4ad+8ah+12am+4ay=4a\times(d+2h+3m+y)$

Now if the whole quantity be divided by one of these factors, according to Art. 118, the quotient will be the other factor.

Thus $(ab+ad)\div a=a+d$. And $(ab+ad)\div(a+d)=a$. Hence,

If the divisor is contained in *every* term of a *compound* div-idend, *it must be cancelled in each.*

Div.	$ab+ac$	$bdk+bdy$	$aah+ay$	$drx+dhx+dxy$
By	a	bd	a	dx
Quot.	$b+c$		$ah+y$	

And if there are *co-efficients*, these must be divided, in each term also.

Div.	$6ab+12ac$	$10dry+16d$	$12hx+8$	$35dm+14dx$
By	$3a$	$2d$	4	$7d$
Quot.	$2b+4c$		$3hx+2$	

121. On the other hand, *if a compound expression contain-ing any factor in every term, be divided by the other quantities connected by their signs, the quotient will be that factor.* See the first part of the preceding article.

Div.	$ab+ac+ah$	$amh+amx+amy$	$4ab+8ay$	$ahm+ahy$
By	$b+c+h$	$h+x+y$	$b+2y$	$m+y$
Quot.	a		$4a$	

122. In division, as well as in multiplication, the caution must be observed, not to confound *terms* with *factors.* See Arts. 99 and 102.

Thus $(ab+ac) \div a = b+c.$ (Art. 120.)
But $(ab \times ac) \div a = aabc \div a = abc.$
And $(ab+ac) \div (b+c) = a.$ (Art. 121.)
But $(ab \times ac) \div (b \times c) = aabc \div bc = aa.$

123. In division the same rule is to be observed respecting the *signs*, as in multiplication; that is, if the *divisor* and *div-idend* are both positive, or both negative, the quotient must be positive: *if one is positive and the other negative, the quotient must be negative.* (Art. 105.)

This is manifest from the consideration that the product of the divisor and quotient must be the same as the dividend.

$$\text{If } \left.\begin{array}{l} +a \times +b = +ab \\ -a \times +b = -ab \\ +a \times -b = -ab \\ -a \times -b = +ab \end{array}\right\} \text{ then } \left\{\begin{array}{l} +ab \div +b = +a \\ -ab \div +b = -a \\ -ab \div -b = +a \\ +ab \div -b = -a \end{array}\right.$$

Div.	abx	$8a-10ay$	$3ax-6ay$	$6am \times dh$
By	$-a$	$-2a$	$3a$	$-2a$
Quot.	$-bx$	$-4+5y$		$-3m \times dh = -3dhm$

124. *If the letters of the divisor are not to be found in the dividend, the division is expressed by writing the divisor under the dividend, in the form of a vulgar fraction.*

Div.	xy	$6hr$	$d-x$	$2d-r$	$d-h+3y$	$r+x+1$
By	a	$4dy$	$-h$	$-3x$	$m-x$	$d+2h-y$
Quot.	$\dfrac{xy}{a}$,		$\dfrac{d-x}{-h}$		$\dfrac{d-h+3y}{m-x}$	

This is a method of *denoting* division, rather than an actual performing of the operation. But the *purposes* of division may frequently be answered, by these fractional expressions. As they are of the same nature with other vulgar fractions, they may be added, subtracted, multiplied, &c. See the next section.

125. When the dividend is a compound quantity, the divisor may either be placed under the *whole* dividend, as in the preceding instances, or it may be repeated under *each term*, taken separately. There are occasions when it will be convenient to exchange one of these forms of expression for the other.

Thus $b+c$ divided by x, is either $\dfrac{b+c}{x}$, or $\dfrac{b}{x}+\dfrac{c}{x}$.

And $a+b$ divided by 2, is either $\dfrac{a+b}{2}$ that is, half the sum of a and b; or $\dfrac{a}{2}+\dfrac{b}{2}$ that is, the sum of half a and half b. For it is evident that *half the sum* of two or more quantities, is equal to the *sum of their halves*. And the same principle is applicable, to a third, fourth, fifth, or any other portion of the dividend.

So also $a-b$ divided by 2, is either $\dfrac{a-b}{2}$, or $\dfrac{a}{2}-\dfrac{b}{2}$.

For *half the difference* of two quantities, is equal to the *difference of their halves.*

So $\dfrac{a-2b+h}{m}=\dfrac{a}{m}-\dfrac{2b}{m}+\dfrac{h}{m}$. And $\dfrac{3a-c}{-x}=\dfrac{3a}{-x}-\dfrac{c}{-x}$.

126. If *some* of the letters in the divisor are in each term of the dividend, the fractional expression may be rendered more simple, by rejecting equal factors from the numerator and denominator.

Div.	ab	dhx	$ahm-3ay$	$ab+bx$	$2am$
By	ac	dy	ab	by	$2xy$
Quot.	$\dfrac{ab}{ac}$ or $\dfrac{b}{c}$		$\dfrac{hm-3y}{b}$		$\dfrac{am}{xy}$

These reductions are made upon the principle, that a given divisor is contained in a given dividend, just as many times, as double the divisor in double the dividend; triple the divisor in triple the dividend, &c. See the reduction of fractions.

127. If the divisor is in some of the terms of the dividend, but not in all; those which contain the divisor may be divided as in Art. 116, and the others set down in the form of a fraction.

Thus $(ab+d)\div a$ is either $\dfrac{ab+d}{a}$, or $\dfrac{ab}{a}+\dfrac{d}{a}$ or $b+\dfrac{d}{a}$.

Div.	$dxy+rx-hd$	$2ab+ad+x$	$bm+3y$	$2my+dh$
By	x	a	$-b$	$2m$
Quot.	$dy+r-\dfrac{hd}{x}$		$-m+\dfrac{3y}{-b}$	

In the four last articles, more than usual caution will be requisite, in applying the *signs.* On this subject, see the next section.

128. The quotient of any quantity divided by *itself* or *its equal,* is obviously *a unit.*

Thus $\dfrac{a}{a}=1$. And $\dfrac{3ax}{3ax}=1$. And $\dfrac{6}{4+2}=1$. And $\dfrac{a+b-3h}{a+b-3h}=1$.

Div.	$ax+x$	$3bd-3d$	$4axy-4a+8ad$	$3ab+3-6m$
By	x	$3d$	$4a$	3

Quot.	$a+1$		$xy-1+2d$	

Cor. If the dividend is *greater* than the divisor, the quotient must be *greater than a unit:* But if the dividend is *less* than the divisor, the quotient must be *less than a unit.*

PROMISCUOUS EXAMPLES.

1. Divide $12aby+6abx-18bbm+24b$, by $6b$.
2. Divide $16a-12+8y+4-20adx+m$, by 4.
3. Divide $(a-2h)\times(3m+y)\times x$, by $(a-2h)\times(3m+y)$.
4. Divide $ahd-4ad+3ay-a$, by $hd-4d+3y-1$.
5. Divide $ax-ry+ad-4my-6+a$, by $-a$.
6. Divide $amy+3my-mxy+am-d$, by $-dmy$.

129. From the nature of division it is evident, that the value of the quotient depends both on the divisor and the dividend. With a given divisor, the greater the dividend, the greater the quotient. And with a given dividend, the greater the divisor, the less the quotient. In several of the succeeding parts of algebra, particularly the subjects of fractions, ratios, and proportion, it will be important to be able to determine what change will be produced in the quotient, by increasing or diminishing either the divisor or the dividend.

If the given dividend be 24, and the divisor 6; the quotient will be 4. But this same dividend may be supposed to be multiplied or divided by some *other* number, before it is divided by 6. Or the *divisor* may be multiplied or divided by some other number, before it is used in dividing 24. In each of these cases, the quotient will be altered.

130. In the first place, if the given divisor is contained in the given dividend a certain number of times, it is obvious that the same divisor is contained,

In *double* that dividend, *twice* as many times;

In *triple* the dividend, *thrice* as many times, &c.

I

That is, if the divisor remains the same, *multiplying the dividend* by any quantity, is, in effect, *multiplying the quotient* by that quantity.

Thus, if the constant divisor is 6, then $\frac{24}{6}=4$ the quotient.

Multiplying the dividend by 2, $\qquad \frac{2\times24}{6}=2\times4$

Multiplying the dividend by 3, $\qquad \frac{3\times24}{6}=3\times4$

Multiplying by any number n, $\qquad \frac{n\times24}{6}=n\times4.$

131. Secondly, if the given divisor is contained in the given dividend a certain number of times, the same divisor is contained,

In *half* that dividend, half as many times;

In *one third* of the dividend, one third as many times, &c.

That is, if the divisor remains the same, *dividing the dividend* by any other quantity, is, in effect, *dividing the quotient* by that quantity.

Thus $\qquad \frac{24}{6}=4.$

Dividing the dividend by 2, $\qquad \frac{\frac{1}{2}24}{6}=\frac{1}{2}4.$

Dividing by 3, $\qquad \frac{\frac{1}{3}24}{6}=\frac{1}{3}4,$ &c.

132. Thirdly, if the given divisor is contained in the given dividend a certain number of times, then, in the same dividend,

Twice that divisor is contained only *half* as many times;

Three times the divisor is contained, *one third* as many times.

That is, if the dividend remains the same, *multiplying the divisor* by any quantity, is, in effect, *dividing the quotient* by that quantity.

Thus $\qquad \frac{24}{6}=4$

Multiplying the divisor by 2, $\qquad \frac{24}{2\times6}=\frac{4}{2}$

Multiplying by 3, $\qquad \frac{24}{3\times6}=\frac{4}{3}$ &c.

133. Lastly, if the given divisor is contained in the given

dividend a certain number of times, then, in the same divi-dend,

Half that divisor is contained, *twice* as many times;

One third of the divisor is contained *thrice* as many times.

That is, if the dividend remains the same, *dividing the divisor* by any other quantity, is, in effect, *multiplying the quotient* by that quantity.

Thus
$$\frac{24}{6}=4.$$

Dividing the divisor by 2,
$$\frac{24}{\frac{1}{2}6}=2\times4$$

Dividing by 3,
$$\frac{24}{\frac{1}{3}6}=3\times4$$

For the method of performing division, when the divisor and dividend are *both compound quantities*, see one of the following sections.

FRACTIONS.*

ART. 134. EXPRESSIONS in the form of fractions occur more frequently in algebra than in arithmetic. Most instances in division belong to this class. Indeed the numerator of *every* fraction may be considered as a *dividend*, of which the denominator is a *divisor*.

According to the common definition in arithmetic, the denominator shows into what parts an integral unit is supposed to be divided; and the numerator shows how many of these parts belong to the fraction. But it makes no difference, whether the *whole* of the numerator is divided by the denominator; or only *one* of the integral units is divided, and then the quotient taken as many times, as the number of units in the numerator. Thus $\frac{3}{4}$ is the same as $\frac{1}{4}+\frac{1}{4}+\frac{1}{4}$. A fourth part of *three* dollars, is equal to three fourths of *one* dollar.

135. The *value* of a fraction, is the *quotient* of the numerator divided by the denominator.

Thus the value of $\frac{6}{2}$ is 3. The value of $\frac{ab}{b}$ is a.

From this it is evident, that whatever changes are made in the *terms* of a fraction; if the *quotient* is not altered, the value remains the same. For any fraction therefore, we may substitute any *other* fraction which will give the same quotient.

Thus $\frac{4}{2}=\frac{10}{5}=\frac{4ba}{2ba}=\frac{8drx}{4drx}=\frac{6+2}{3+1}$ &c. For the quotient in each of these instances is 2.

136. As the value of a fraction is the quotient of the numerator divided by the denominator, it is evident, from Art. 128, that when the numerator is *equal* to the denominator, the value of the fraction is *a unit*; when the numerator is

* Horsley's Mathematics, Camus' Arithmetic, Emerson, Euler, Saunderson, and Ludlam.

less than the denominator, the value is *less than a unit ;* and when the numerator is *greater* than the denominator, the value is *greater than a unit.*

The calculations in fractions depend on a few general principles, which will here be stated in connection with each other.

137. *If the denominator of a fraction remains the same, multiplying the* NUMERATOR *by any quantity, is multiplying the* VALUE *by that quantity ; and dividing the numerator, is dividing the value.* For the numerator and denominator are a dividend and divisor, of which the value of the fraction is the quotient. And by Art. 130 and 131, multiplying the dividend is in effect multiplying the quotient, and dividing the dividend is dividing the quotient.

Thus, in the fractions $\dfrac{ab}{a}, \dfrac{3ab}{a}, \dfrac{7abd}{a}, \dfrac{\frac{1}{3}ab}{a}, \dfrac{\frac{1}{4}abd}{a}$, &c.

The quot's or values are b, $3b$, $7bd$, $\frac{1}{3}b$, $\frac{1}{4}bd$, &c.

Here it will be seen that, while the denominator is not altered, the value of the fraction is multiplied or divided by the same quantity as the numerator.

Cor. With a given denominator, the greater the numerator, the greater will be the *value* of the fraction ; and, on the other hand, the greater the value, the greater the numerator.

138. *If the numerator remains the same, multiplying the denominator by any quantity, is dividing the value by that quantity ; and dividing the denominator, is multiplying the value.* For multiplying the divisor is dividing the quotient ; and dividing the divisor is multiplying the quotient. (Art. 132, 133.)

In the fractions $\dfrac{24ab}{6b}, \dfrac{24ab}{12b}, \dfrac{24ab}{3b}, \dfrac{24ab}{b}$, &c.

The values are $4a$, $2a$, $8a$, $24a$, &c.

Cor. With a given numerator, the *greater* the denominator, the *less* will be the value of the fraction ; and the less the value, the greater the denominator

139. From the two last articles it follows, that *dividing the numerator* by any quantity, will have the same effect on the value of the fraction, as *multiplying the denominator* by that quantity ; and *multiplying the numerator* will have the same effect, as *dividing the denominator.*

140. It is also evident, from the preceding articles, that *if the numerator and denominator be both multiplied, or both*

divided, by the same quantity, the value of the fraction will not be altered.

Thus $\dfrac{bx}{b} = \dfrac{abx}{ab} = \dfrac{3bx}{3b} = \dfrac{\frac{1}{2}bx}{\frac{1}{2}b} = \dfrac{\frac{1}{3}abx}{\frac{1}{3}ab}$ &c. For in each of these instances the quotient is x.

141. Any integral quantity may, without altering its value, be thrown into the form of a fraction, by multiplying the quantity into the proposed denominator, and taking the product for a numerator.

Thus $a = \dfrac{a}{1} = \dfrac{ab}{b} = \dfrac{ad+ah}{d+h} = \dfrac{6adh}{6dh}$, &c. For the quotient of each of these is a.

So $d+h = \dfrac{dx+hx}{x}$. And $r+1 = \dfrac{2drr+2dr}{2dr}$.

142. There is nothing perhaps, in the calculation of algebraic fractions, which occasions more perplexity to a learner, than the positive and negative *signs*. The changes in these are so frequent, that it is necessary to become familiar with the principles on which they are made. The use of the sign which is prefixed to the dividing line, is to show whether the value of *the whole fraction* is to be added to, or subtracted from, the other quantities with which it is connected. (Art. 43.) This sign, therefore, has an influence on the several terms taken collectively. But in the numerator and denominator, each sign affects only the single term to which it is applied.

The value of $\dfrac{ab}{b}$ is a. (Art. 135.) But this will become negative, if the sign — be prefixed to the fraction.

Thus $y + \dfrac{ab}{b} = y + a$. But $y - \dfrac{ab}{b} = y - a$.

So that changing the sign which is before the whole fraction, has the effect of changing the *value* from positive to negative, or from negative to positive.

Next, suppose the sign or signs of the *numerator* to be changed.

By Art. 123, $\dfrac{ab}{b} = +a$. But $\dfrac{-ab}{b} = -a$.

And $\dfrac{ab-bc}{b} = +a-c$. But $\dfrac{-ab+bc}{b} = -a+c$.

That is, by changing all the signs of the numerator, the

value of the fraction is changed from positive to negative, or the contrary.

Again, suppose the sign of the *denominator* to be changed.

$$\text{As before } \frac{ab}{b} = +a. \qquad \text{But } \frac{ab}{-b} = -a.$$

. 143. We have, then, this general proposition; *If the sign prefixed to a fraction, or all the signs of the numerator, or all the signs of the denominator be changed; the value of the fraction will be changed, from positive to negative, or from negative to positive.*

From this is derived another important principle. As each of the changes mentioned here is from positive to negative, or the contrary; if any *two* of them be made at the same time, *they will balance each other.*

Thus, by changing the sign of the numerator,

$$\frac{ab}{b} = +a \text{ becomes } \frac{-ab}{b} = -a.$$

But, by changing both the numerator and denominator, it becomes $\frac{-ab}{-b} = +a$, where the positive value is restored.

By changing the sign before the fraction,

$$y + \frac{ab}{b} = y + a \text{ becomes } y - \frac{ab}{b} = y - a.$$

But, by changing the sign of the numerator also, it becomes $y - \frac{-ab}{b}$ where the quotient $-a$ is to be *subtracted* from y, or which is the same thing, (Art. 81,) $+a$ is to be *added*, making $y + a$ as at first. Hence,

144. *If all the signs both of the numerator and denominator, or the signs of one of these with the sign prefixed to the whole fraction, be changed at the same time, the value of the fraction will not be altered.*

$$\text{Thus } \frac{6}{2} = \frac{-6}{-2} = -\frac{-6}{2} = -\frac{6}{-2} = +3.$$

$$\text{And } \frac{6}{-2} = \frac{-6}{2} = -\frac{6}{2} = -\frac{-6}{-2} = -3.$$

Hence the quotient in division may be set down in different ways. Thus $(a-c) \div b$, is either $\frac{a}{b} + \frac{-c}{b}$, or $\frac{a}{b} - \frac{c}{b}$.

The latter method is the most common. See the examples in Art. 127.

Reduction of Fractions.

145. From the principles which have been stated, are derived the rules for the *Reduction* of fractions, which are substantially the same in algebra, as in arithmetic.

A fraction may be reduced to lower terms, by dividing both the numerator and denominator, by any quantity which will divide them without a remainder. According to Art. 140, this will not alter the *value* of the fraction.

Thus $\dfrac{ab}{cb}=\dfrac{a}{c}$. And $\dfrac{6dm}{8dy}=\dfrac{3m}{4y}$. And $\dfrac{7m}{7mr}=\dfrac{1}{r}$.

In the last example, both parts of the fraction are divided by the numerator. The reduced numerator must therefore be a unit. (Art. 128.)

Again $\dfrac{a+bc}{(a+bc)\times m}=\dfrac{1}{m}$ (Art. 118.) And $\dfrac{am+ay}{bm+by}=\dfrac{a}{b}$. (Art. [121.) (Art.

If a letter is in *every* term both of the numerator and denominator, it may be *cancelled*, for this is *dividing* by that letter. (Art. 120.)

Thus $\dfrac{3am+ay}{ad+ah}=\dfrac{3m+y}{d+h}$ And $\dfrac{dry+dy}{dhy-dy}=\dfrac{r+1}{h-1}$.

146. *Fractions of different denominators may be reduced to a common denominator, by multiplying each numerator into all the denominators except its own, for a new numerator; and all the denominators together, for a common denominator.*

Ex. 1. Reduce $\dfrac{a}{b}$, and $\dfrac{c}{d}$, and $\dfrac{m}{y}$ to a common denominator. .

$\left.\begin{array}{l} a\times d\times y=ady \\ c\times b\times y=cby \\ m\times b\times d=mbd \end{array}\right\}$ the three denominators.

$b\times d\times y=bdy$ the common denominator,

The fractions reduced are $\dfrac{ady}{bdy}$, and $\dfrac{bcy}{bdy}$, and $\dfrac{bdm}{bdy}$.

Here it will be seen, that the reduction consists in multiplying the numerator and denominator of each fraction, into all the other denominators. This does not alter the value. (Art. 140.)

2. Reduce $\dfrac{dr}{3m}$, and $\dfrac{2h}{g}$, and $\dfrac{6c}{y}$.

Ans. $\dfrac{dgry}{3gmy}$, and $\dfrac{6hmy}{3gmy}$, and $\dfrac{18cgm}{3gmy}$.

3. Reduce $\dfrac{2}{3}$, and $\dfrac{a}{x}$, and $\dfrac{r+1}{d+h}$.

Ans. $\dfrac{2dx+2hx}{3dx+3hx}$, and $\dfrac{3ad+3ah}{3dx+3hx}$, and $\dfrac{3rx+3x}{3dx+3hx}$.

4. Reduce $\dfrac{1}{a+b}$, and $\dfrac{1}{a-b}$.

Ans. $\dfrac{a-b}{aa-bb}$, and $\dfrac{a+b}{aa-bb}$.

After the fractions have been reduced to a common denominator, they may be reduced to lower terms, by the rule in the last article, if there is any quantity, which will divide the denominator, and *all* the numerators, without a remainder.

An *integer* and a fraction are easily reduced to a common denominator. [(Art. 141.)

Thus a and $\dfrac{b}{c}$ are equal to $\dfrac{a}{1}$ and $\dfrac{b}{c}$, or $\dfrac{ac}{c}$ and $\dfrac{b}{c}$.

And a, b, $\dfrac{h}{m}$, $\dfrac{d}{y}$ are equal to $\dfrac{amy}{my}$, $\dfrac{bmy}{my}$, $\dfrac{hy}{my}$, $\dfrac{dm}{my}$.

147. *To reduce an improper fraction to a mixed quantity, divide the numerator by the denominator, as in* Art. 127.

Thus $\dfrac{ab+bm+d}{b}=a+m+\dfrac{d}{b}$.

And $\dfrac{dx+d-7h+y}{d}=x+1-\dfrac{7h}{d}+\dfrac{y}{d}$.

Reduce $\dfrac{am-a+ady-hr}{a}$, to a mixed quantity.

For the reduction of a *mixed quantity* to an improper fraction, see Art. 150. And for the reduction of a *compound fraction* to a simple one, see Art. 160.

J

ADDITION OF FRACTIONS.

148. In adding fractions, we may either write them one after the other, with their signs, as in the addition of integers, or we may incorporate them into a single fraction, by the following rule :

Reduce the fractions to a common denominator, make the signs before them all positive, and then add their numerators.

The common denominator shows into what parts the integral unit is supposed to be divided; and the numerators show the *number* of these parts belonging to each of the fractions. (Art. 134.) Therefore the numerators *taken together* show the whole number of parts in all the fractions.

Thus $\frac{2}{7} = \frac{1}{7} + \frac{1}{7}$. And $\frac{3}{7} = \frac{1}{7} + \frac{1}{7} + \frac{1}{7}$,

Therefore $\frac{2}{7} + \frac{3}{7} = \frac{1}{7} + \frac{1}{7} + \frac{1}{7} + \frac{1}{7} + \frac{1}{7} = \frac{5}{7}$.

The numerators are added, according to the rules for the addition of integers. (Art 69, &c.) It is obvious that the sum is to be placed over the common denominator. To avoid the perplexity which might be occasioned by the signs, it will be expedient to make those prefixed to the fractions uniformly positive. But in doing this, care must be taken not to alter the value. This will be preserved, if all the signs in the numerator, are changed at the same time with that before the fraction. (Art. 144.)

Ex. 1. Add $\frac{2}{16}$ and $\frac{4}{16}$ of a pound. Ans. $\frac{2+4}{16}$ or $\frac{6}{16}$.

It is as evident that $\frac{2}{16}$, and $\frac{4}{16}$ of a pound, are $\frac{6}{16}$ of a pound, as that 2 ounces, and 4 ounces, are 6 ounces.

2. Add $\frac{a}{b}$ and $\frac{c}{d}$. First reduce them to a common denominator. They will then be $\frac{ad}{bd}$ and $\frac{bc}{bd}$, and their sum $\frac{ad+bc}{bd}$.

3. Given $\dfrac{m}{d}$ and $-\dfrac{2r+d}{3h}$, to find their sum.

Ans, $\dfrac{m}{d}$ and $-\dfrac{2r+d}{3h}=\dfrac{3hm}{3dh}$ and $-\dfrac{2dr+dd}{3dh}=\dfrac{3hm-2dr-dd}{3dh}$. Art. [144.

4. $\dfrac{a}{d}$ and $-\dfrac{b-m}{y}=\dfrac{a}{d}+\dfrac{-b+m}{y}=\dfrac{ay-bd+dm}{dy}$.

5. $\dfrac{a}{y}$ and $\dfrac{d}{-m}=\dfrac{-am}{-my}+\dfrac{dy}{-my}=\dfrac{-am+dy}{-my}$ or $\dfrac{am-dy}{my}$. (Art. [144.)

6. $\dfrac{a}{a+b}$ and $\dfrac{b}{a-b}=\dfrac{aa-ab+ab+bb}{aa+ab-ab-bb}=\dfrac{aa+bb}{aa-bb}$. (Art. 77.)

7. Add $\dfrac{-a}{d}$ to $\dfrac{-h}{m-r}$.   8. Add $\dfrac{-4}{2}$ to $\dfrac{-16}{7-3}$. Ans. -6.

149. For many purposes, it is sufficient to add fractions in the same manner as integers are added, by writing them one after another with their signs. (Art. 69.)

Thus the sum of $\dfrac{a}{b}$ and $\dfrac{3}{y}$ and $-\dfrac{d}{2m}$, is $\dfrac{a}{b}+\dfrac{3}{y}-\dfrac{d}{2m}$.

In the same manner, fractions and integers may be added.

The sum of a and $\dfrac{d}{y}$ and $3m$ and $-\dfrac{h}{r}$, is $a+3m+\dfrac{d}{y}-\dfrac{h}{r}$.

150. Or the integer may be *incorporated* with the fraction, by giving to the former the denominator of the latter, and then adding the numerators. See Art. 141.

The sum of a and $\dfrac{b}{m}$, is $\dfrac{a}{1}+\dfrac{b}{m}=\dfrac{am}{m}+\dfrac{b}{m}=\dfrac{am+b}{m}$.

The sum of $3d$ and $\dfrac{h+d}{m-y}$, is $\dfrac{3dm-3dy+h+d}{m-y}$.

Incorporating an integer with a fraction, is the same as *reducing a mixed quantity* to an improper fraction. For a mixed quantity is an integer and a fraction. In arithmetic, these are generally placed together, without any sign between them. But in algebra, they are distinct terms. Thus $2\frac{1}{3}$ is 2 *and* $\frac{1}{3}$, which is the same as $2+\frac{1}{3}$.

Ex. 1. Reduce $a+\frac{1}{b}$ to an improper fraction. Ans. $\frac{ab+1}{b}$.

2. Reduce $6\frac{3}{4}$. Ans. $6\frac{3}{4}=6+\frac{3}{4}=\frac{24+3}{4}=\frac{27}{4}$.

3. Reduce $m+d-\frac{r}{h-d}$. Ans. $\frac{hm-dm+dh-dd-r}{h-d}$,

4. Reduce $1+\frac{d}{b}$. Ans. $\frac{b+d}{b}$.   5. Reduce $1-\frac{h}{m}$.

6. Reduce $b+\frac{c}{d-y}$.   7. Reduce $3+\frac{2d-4}{3a}$.

Subtraction of Fractions.

151. The methods of performing subtraction in algebra, depend on the principle, that adding a negative quantity is equivalent to subtracting a positive one; and *v. v.* (Art. 81.) For the subtraction of fractions, then, we have the following simple rule. *Change the fraction to be subtracted, from positive to negative, or the contrary, and then proceed as in addition.* (Art. 148.) In making the required change, it will be expedient to alter, in some instances the signs of the numerator, and in others, the sign before the dividing line, (Art. 143,) so as to leave the latter always affirmative.

Ex. 1. From $\frac{a}{b}$, subtract $\frac{h}{m}$.

First change $\frac{h}{m}$, the fraction to be subtracted, to $\frac{-h}{m}$.

Secondly, reduce the two fractions to a common denominator, making $\frac{am}{bm}$ and $\frac{-bh}{bm}$.

Thirdly, take the sum of the numerators, $am-bh$:

This, placed over the common denominator, gives the answer required, $\frac{am-bh}{bm}$.

2. From $\dfrac{a+y}{r}$, subtract $\dfrac{h}{d}$.　Ans. $\dfrac{ad+dy-hr}{dr}$.

3. From $\dfrac{a}{m}$ subtract $\dfrac{d-b}{y}$.　Ans. $\dfrac{ay-dm+bm}{my}$.

4. From $\dfrac{a+3d}{4}$, subtract $\dfrac{3a-2d}{3}$.　Ans. $\dfrac{17d-9a}{12}$.

5. From $\dfrac{b-d}{m}$ subtract $-\dfrac{b}{y}$.　Ans. $\dfrac{b-d}{m}+\dfrac{b}{y}=\dfrac{by-dy+bm}{my}$.

6. From $\dfrac{a+1}{d}$ subtract $\dfrac{d-1}{m}$.　　7. From $\dfrac{3}{a}$ subtract $\dfrac{4}{b}$.

152. Fractions may also be subtracted, like integers, by setting them down, after their signs are changed, without reducing them to a common denominator.

Ex. 1. From $\dfrac{a}{b}$ subtract $\dfrac{d}{h}$.　Ans. $\dfrac{a}{b}-\dfrac{d}{h}$.

2. From $\dfrac{h}{m}$ subtract $-\dfrac{h+d}{y}$.　Ans. $\dfrac{h}{m}+\dfrac{h+d}{y}$.

In the same manner, an integer may be subtracted from a fraction, or a fraction from an integer.

3. From a subtract $\dfrac{b}{m}$.　Ans. $a-\dfrac{b}{m}$.

153. Or the integer may be incorporated with the fraction, as in Art. 150.

Ex. 1. From $\dfrac{h}{y}$ subtract m.　Ans. $\dfrac{h}{y}-m=\dfrac{h-my}{y}$.

2. From $4a+\dfrac{b}{c}$ subtract $3a-\dfrac{h}{d}$.

Ans. $a+\dfrac{bd+hc}{cd}=\dfrac{acd+bd+hc}{cd}$.

3. From $1+\dfrac{b-c}{d}$ subtract $\dfrac{c-b}{d}$. Ans. $\dfrac{d+2b-2c}{d}$.

4. From $a+3h-\dfrac{d-b}{2}$ subtract $3a-h+\dfrac{d+b}{3}$.

MULTIPLICATION OF FRACTIONS.

154. By the definition of multiplication, multiplying by a fraction is taking a *part* of the multiplicand, as many times, as there are like parts of an unit in the multiplier. (Art. 90.) Now the denominator of a fraction shows into what parts the integral unit is supposed to be divided; and the numerator shows how many of those parts belong to the given fraction. In multiplying by a fraction, therefore, the multiplicand is to be divided into such parts, as are denoted by the denominator; and then one of these parts is to be repeated, as many times, as is required by the numerator.

Suppose a is to be multiplied by $\dfrac{3}{4}$.

A fourth part of a is $\qquad$ $\dfrac{a}{4}$.

This taken 3 times is $\qquad$ $\dfrac{a}{4}+\dfrac{a}{4}+\dfrac{a}{4}=\dfrac{3a}{4}$. (Art. 148.)

Again, suppose $\dfrac{a}{b}$ is to be multiplied by $\dfrac{3}{4}$.

One fourth of $\dfrac{a}{b}$ is $\qquad$ $\dfrac{a}{4b}$. (Art. 138.)

This taken 3 times is $\qquad$ $\dfrac{a}{4b}+\dfrac{a}{4b}+\dfrac{a}{4b}=\dfrac{3a}{4b}$

the product required.

In a similar manner, any fractional multiplicand may be divided into parts, by multiplying the denominator; and one of the parts may be repeated, by multiplying the numerator. We have then the following rule :

155. *To multiply fractions, multiply the numerators togeth-*
er, for a new numerator, and the denominators together, for
a new denominator.

Ex. 1. Multiply $\dfrac{3b}{c}$ into $\dfrac{d}{2m}$. Product $\dfrac{3bd}{2cm}$.

2. Multiply $\dfrac{a+d}{y}$ into $\dfrac{4h}{m-2}$. Product $\dfrac{4ah+4dh}{my-2y}$

3. Mult. $\dfrac{(a+m)\times h}{3}$ into $\dfrac{4}{(a-n)}$. Prod. $\dfrac{(a+m)\times 4h}{3\times(a-n)}$. (Art. 99.)

4. Mult. $\dfrac{a+b+3h}{y}$ into $\dfrac{1}{d+1}$. Prod. $\dfrac{a+b+3h}{dy+y}$.

5. Mult. $\dfrac{a+h}{3+d}$ into $\dfrac{4-m}{c+y}$. 6. Mult. $\dfrac{1}{a-3r}$ into $\dfrac{3}{8}$.

156. The method of multiplying is the same, when there
are more than two fractions to be multiplied together.

1. Multiply together $\dfrac{a}{b}$, $\dfrac{c}{d}$, and $\dfrac{m}{y}$. Product $\dfrac{acm}{bdy}$.

For $\dfrac{a}{b}\times\dfrac{c}{d}$ is, by the last article $\dfrac{ac}{bd}$, and this into $\dfrac{m}{y}$, is $\dfrac{acm}{bdy}$.

2. Multiply $\dfrac{2a}{m}$, $\dfrac{h-d}{y}$, $\dfrac{b}{c}$, and $\dfrac{1}{r-1}$. Product $\dfrac{2abh-2abd}{cmry-cmy}$

3. Mult. $\dfrac{3+b}{n}$, $\dfrac{1}{h}$, and $\dfrac{d}{r+2}$. 4. Mult. $\dfrac{ad}{ky}$, $\dfrac{a-6}{d+1}$, and $\dfrac{3}{7}$.

157. The multiplication may sometimes be shortened, by
rejecting equal factors, from the numerators and denominators.

1. Multiply $\dfrac{a}{r}$ into $\dfrac{h}{a}$ and $\dfrac{d}{y}$. Product $\dfrac{dh}{ry}$.

Here a being in one of the numerators, and in one of the

denominators, may be omitted. If it be retained, the product will be $\dfrac{adh}{ary}$. But this reduced to lower terms, by Art. 145, will become $\dfrac{dh}{ry}$ as before.

2. Mult. $\dfrac{ad}{m}$ into $\dfrac{m}{3a}$ and $\dfrac{ah}{2d}$. Product $\dfrac{ah}{6}$.

It is necessary that the factors rejected from the numerators be exactly equal to those which are rejected from the denominators. In the last example, a being in two of the numerators, and in only one of the denominators, must be retained in one of the numerators.

3. Mult. $\dfrac{a+d}{y}$ into $\dfrac{my}{ah}$. Prod. $\dfrac{am+dm}{ah}$.

Here, though the same letter a is in one of the numerators, and in one of the denominators, yet as it is not in *every term* of the numerator, it must not be cancelled.

4. Mult. $\dfrac{am+d}{h}$ into $\dfrac{h}{m}$ and $\dfrac{3r}{5a}$. Prod. $\dfrac{3amr+3dr}{5am}$.

If any difficulty is found, in making these contractions, it will be better to perform the multiplication, without omitting any of the factors; and to reduce the product to lower terms afterwards.

158. When a fraction and an *integer* are multiplied together, the *numerator* of the fraction is multiplied into the integer. The denominator is not altered; except in cases where division of the denominator is substituted for multiplication of the numerator, according to Art. 139.

Thus $a \times \dfrac{m}{y} = \dfrac{am}{y}$. For $a = \dfrac{a}{1}$; and $\dfrac{a}{1} \times \dfrac{m}{y} = \dfrac{am}{y}$.

So $r \times \dfrac{x}{d} \times \dfrac{h+1}{3} = \dfrac{hrx+rx}{3d}$. And $a \times \dfrac{1}{b} = \dfrac{a}{b}$. Hence,

159. *A fraction is multiplied into a quantity equal to its de-nominator, by cancelling the denominator.*

Thus $\frac{a}{b} \times b = a.$ For $\frac{a}{b} \times b = \frac{ab}{b}.$ But the letter b, be-ing in both the numerator and denominator, may be set aside. (Art. 145.)

So $\frac{3m}{a-y} \times (a-y) = 3m.$ And $\frac{h+3d}{3+m} \times (3+m) = h+3d.$

On the same principle, a fraction is multiplied into any *factor* in its denominator, by cancelling that factor.

Thus $\frac{a}{by} \times y = \frac{ay}{by} = \frac{a}{b}.$ And $\frac{h}{24} \times 6 = \frac{h}{4}.$

160. From the definition of multiplication by a fraction, it follows that what is commonly called a *compound fraction,*[*] is the *product* of two or more fractions. Thus $\frac{3}{4}$ of $\frac{a}{b}$ is

$\frac{3}{4} \times \frac{a}{b}.$ For $\frac{3}{4}$ of $\frac{a}{b}$, is $\frac{1}{4}$ of $\frac{a}{b}$ taken three times, that is,

$\frac{a}{4b} + \frac{a}{4b} + \frac{a}{4b}.$ But this is the same as $\frac{a}{b}$ multiplied by $\frac{3}{4}.$ (Art. 154.)

Hence, *reducing a compound fraction to a simple one, is the same, as multiplying fractions into each other.*

Ex. 1. Reduce $\frac{2}{7}$ of $\frac{a}{b+2}.$ Ans. $\frac{2a}{7b+14}.$

2. Reduce $\frac{2}{3}$ of $\frac{4}{5}$ of $\frac{b+h}{2a-m}.$ Ans. $\frac{8b+8h}{30a-15m}.$

3. Reduce $\frac{1}{7}$ of $\frac{1}{3}$ of $\frac{1}{8-d}.$ Ans. $\frac{1}{168-21d}.$

161. The expressions $\frac{2}{3}a$, $\frac{1}{5}b$, $\frac{4}{7}y$, &c. are equivalent to $\frac{2a}{3}$, $\frac{b}{5}$, $\frac{4y}{7}$, &c. For $\frac{2}{3}a$ is $\frac{2}{3}$ of a, which is equal to $\frac{2}{3} \times a = \frac{2a}{3}.$ (Art. 158.) So $\frac{1}{5}b = \frac{1}{5} \times b = \frac{b}{5}.$

---

[*] By a compound fraction is meant a fraction *of* a fraction, and not a fraction whose numerator or denominator is a compound quantity.

K

DIVISION OF FRACTIONS.

162. *To divide one fraction by another, invert the divisor, and then proceed as in multiplication.* (Art 155.)

Ex. 1. Divide $\dfrac{a}{b}$ by $\dfrac{c}{d}$. Ans. $\dfrac{a}{b} \times \dfrac{d}{c} = \dfrac{ad}{bc}$.

To understand the reason of the rule, let it be premised, that the product of any fraction into the same fraction inverted is always a unit.

Thus $\dfrac{a}{b} \times \dfrac{b}{a} = \dfrac{ab}{ab} = 1$.   And $\dfrac{d}{h+y} \times \dfrac{h+y}{d} = \dfrac{dh+dy}{dh+dy} = 1$.   (Art. 128.)

But a quantity is not altered by multiplying it by a unit. Therefore if a dividend be multiplied, first into the divisor inverted, and then into the divisor itself, the last product will be equal to the dividend. Now, by the definition, art. 115, "division is finding a quotient, which multiplied into the divisor will produce the dividend." And as the dividend multiplied into the divisor inverted is such a quantity, the quotient is truly found by the rule.

This explanation will probably be best understood, by attending to the examples. In several which follow, tho *proof* of the division will be given, by multiplying the quotient into the divisor. This will present, at one view, the dividend multiplied into the inverted divisor, and into the divisor itself.

2. Divide $\dfrac{m}{2d}$ by $\dfrac{3h}{y}$.   Ans. $\dfrac{m}{2d} \times \dfrac{y}{3h} = \dfrac{my}{6dh}$

Proof. $\dfrac{my}{6dh} \times \dfrac{3h}{y} = \dfrac{3hmy}{6dhy} = \dfrac{m}{2d}$ the dividend.

3. Divide $\dfrac{x+d}{r}$ by $\dfrac{5d}{y}$.   Ans. $\dfrac{x+d}{r} \times \dfrac{y}{5d} = \dfrac{xy+dy}{5dr}$.

Proof. $\dfrac{xy+dy}{5dr} \times \dfrac{5d}{y} = \dfrac{5dxy+5ddy}{5dry} = \dfrac{x+d}{r}$

4. Divide $\dfrac{4dh}{x}$ by $\dfrac{4hr}{a}$.   Ans. $\dfrac{4dh}{x} \times \dfrac{a}{4hr} = \dfrac{ad}{rx}$ (Art.157.)

Proof. $\dfrac{ad}{rx} \times \dfrac{4hr}{a} = \dfrac{4adhr}{arx} = \dfrac{4dh}{x}$ the dividend.

5. Divide $\dfrac{36d}{5}$ by $\dfrac{18h}{10y}$. Ans. $\dfrac{36d}{5} \times \dfrac{10y}{18h} = \dfrac{4dy}{h}$.

Proof $\dfrac{4dy}{h} \times \dfrac{18h}{10y} = \dfrac{36d}{5}$, the dividend.

6. Divide $\dfrac{ab+1}{3y}$ by $\dfrac{ab-1}{x}$.　　7. Divide $\dfrac{h-my}{4}$ by $\dfrac{3}{a+1}$.

163. When a fraction is divided by an *integer*, the *denominator* of the fraction is multiplied into the integer.

Thus the quotient of $\dfrac{a}{b}$ divided by m, is $\dfrac{a}{bm}$.

For $m = \dfrac{m}{1}$; and by the last article, $\dfrac{a}{b} \div \dfrac{m}{1} = \dfrac{a}{b} \times \dfrac{1}{m} = \dfrac{a}{bm}$.

So $\dfrac{1}{a-b} \div h = \dfrac{1}{a-b} \times \dfrac{1}{h} = \dfrac{1}{ah-bh}$. And $\dfrac{3}{4} \div 6 = \dfrac{3}{24} = \dfrac{1}{8}$.

In fractions, multiplication is made to perform the office of division; because division in the usual form often leaves a troublesome remainder. But there is no remainder in multiplication. In many cases, there are methods of shortening the operation. But these will be suggested by practice, without the aid of particular rules.

164. By the definition, art. 49, "the *reciprocal* of a quantity, is the quotient arising from dividing a unit by that quantity."

Therefore, the reciprocal of $\dfrac{a}{b}$, is $1 \div \dfrac{a}{b} = 1 \times \dfrac{b}{a} = \dfrac{b}{a}$. That is,

The reciprocal of a fraction is the fraction inverted.

Thus the reciprocal of $\dfrac{b}{m+y}$ is $\dfrac{m+y}{b}$; the reciprocal of $\dfrac{1}{3y}$ is $\dfrac{3y}{1}$ or $3y$; the reciprocal of $\frac{1}{4}$ is 4. Hence the reciprocal of a fraction whose numerator is 1, is the denominator of the fraction.

Thus the reciprocal of $\dfrac{1}{a}$ is a; of $\dfrac{1}{a+b}$, is $a+b$, &c.

165. A fraction sometimes occurs in the numerator or denominator of another fraction, as $\dfrac{\frac{2}{3}a}{b}$. It is often convenient,

in the course of a calculation, to transfer such a fraction, from the numerator to the denominator of the principal fraction, or the contrary. That this may be done, without altering the value, if the fraction transferred be *inverted*, is evident, from the following principles:

First, *Dividing* by a fraction, is the same as *multiplying* by the fraction *inverted*. (Art. 162.)

Secondly, *Dividing the numerator* of a fraction has the same effect on the value, as *multiplying the denominator;* and multiplying the numerator has the same effect, as dividing the denominator. (Art. 139.)

Thus in the expression $\dfrac{\frac{3}{5}a}{x}$, the numerator of $\dfrac{a}{x}$ is multiplied into $\frac{3}{5}$. But the value will be the same, if, instead of multiplying the numerator, we divide the denominator by $\frac{3}{5}$, that is, multiply the denominator by $\frac{5}{3}$.

Therefore $\dfrac{\frac{3}{5}a}{x}=\dfrac{a}{\frac{5}{3}x}$. $\qquad$ So $\dfrac{h}{\frac{7}{9}m}=\dfrac{\frac{9}{7}h}{m}$.

And $\dfrac{\frac{3}{4}d}{h+y}=\dfrac{d}{\frac{4}{3}\times(h+y)}=\dfrac{d}{\frac{4}{3}h+\frac{4}{3}y}$. And $\dfrac{a-x}{\frac{2}{7}m}=\dfrac{\frac{7}{2}a-\frac{7}{2}x}{m}$.

166. Multiplying the *numerator* is in effect multiplying the *value* of the fraction. (Art. 137.) On this principle, a fraction may be cleared of a fractional co-efficient which occurs in its numerator.

Thus $\dfrac{\frac{3}{5}a}{b}=\dfrac{3}{5}\times\dfrac{a}{b}=\dfrac{3a}{5b}$. And $\dfrac{\frac{1}{5}a}{y}=\dfrac{1}{5}\times\dfrac{a}{y}=\dfrac{a}{5y}$.

And $\dfrac{\frac{1}{3}h+\frac{1}{3}x}{m}=\dfrac{1}{3}\times\dfrac{h+x}{m}=\dfrac{h+x}{3m}$. And $\dfrac{\frac{3}{4}x}{5a}=\dfrac{3x}{20a}$.

On the other hand, $\dfrac{3a}{7x}=\dfrac{3}{7}\times\dfrac{a}{x}=\dfrac{\frac{3}{7}a}{x}$.

And $\dfrac{a}{3y}=\dfrac{1}{3}\times\dfrac{a}{y}=\dfrac{\frac{1}{3}a}{y}$. And $\dfrac{4a}{5d+5x}=\dfrac{\frac{4}{5}a}{d+x}$.

167. But multiplying the *denominator*, by another fraction, is in effect *dividing* the value; (Art. 138.) that is, it is *multiplying* the value by the fraction *inverted*. The principal fraction may therefore be cleared of a fractional co-efficient, which occurs in its denominator.

Thus $\dfrac{a}{\frac{3}{5}b}=\dfrac{a}{b}\div\dfrac{3}{5}=\dfrac{a}{b}\times\dfrac{5}{3}=\dfrac{5a}{3b}$.　And $\dfrac{a}{\frac{2}{7}x}=\dfrac{7a}{2x}$.

And $\dfrac{a+h}{\frac{3}{9}y}=\dfrac{9a+9h}{3y}$.　　　　And $\dfrac{3h}{\frac{4}{7}m}=\dfrac{21h}{4m}$.

On the other hand, $\dfrac{7a}{3x}=\dfrac{a}{\frac{3}{7}x}$.

And $\dfrac{3y+3dx}{2m}=\dfrac{y+dx}{\frac{2}{3}m}$.　　　　　　And $\dfrac{3x}{y}=\dfrac{x}{\frac{1}{3}y}$

SIMPLE EQUATIONS.

ART. 168. THE subjects of the preceding sections are introductory to what may be considered the peculiar province of algebra, the investigation of the values of unknown quantities, by means of *equations.*

An equation is a proposition, expressing in algebraic charac-ters, the equality between one quantity or set of quantities and another. Thus $x+a=b+c$, is an equation, in which the sum of x and a, is equal to the sum of b and c. The quan-tities on the two sides of the sign of equality, are sometimes called the *members* of the equation; the several terms on the *left* constituting the *first* member, and those on the *right*, the *second* member. In the equation $a+y=d-x$, the first member is $a+y$, and the second $d-x$.

169. The object aimed at, in what is called the *resolution* or *reduction* of an equation is to *find the value of the unknown quantity.* In the first statement of the conditions of a prob-lem, the known and unknown quantities are frequently thrown promiscuously together. To find the value of that which is required, it is necessary to bring it to stand by it-self, while all the others are on the opposite side of the equation. But, in doing this, care must be taken not to *destroy* the equation, by rendering the two members une-qual. Many changes may be made in the arrangement of the terms, without affecting the equality of the sides.

170. *The reduction of an equation consists, then, in bring-ing the unknown quantity by itself, on one side, and all the known quantities on the other side, without destroying the equa-tion.*

To effect this, it is evident that one of the members must be as much increased or diminished as the other. If a quan-tity be added to one, and not to the other, the equality will be destroyed. But the members will remain equal;

If the same or equal quantities be *added* to each. Ax. 1.

If the same or equal quantities be *subtracted* from each. Ax. 2.

If each be *multiplied* by the same or equal quantities. Ax. 3.
If each be *divided* by the same or equal quantities. Ax. 4.

171. It may be farther observed that, in general, if the unknown quantity is connected with others by addition, multiplication, division, &c. the reduction is made by a *contrary* process. If a known quantity is *added* to the unknown, the equation is reduced by *subtraction*. If one is *multiplied* by the other, the reduction is effected by *division*, &c. The reason of this will be seen, by attending to the several cases in the following articles. The *known* quantities may be expressed either by letters or figures. The *unknown* quantity is represented by one of the last letters of the alphabet, generally x, y, or z. (Art. 27.) The principal reductions to be considered in this section, are those which are effected by *transposition*, *multiplication*, and *division*. These ought to be made perfectly familiar, as one or more of them will be necessary, in the resolution of almost every equation.

Transposition.

172. In the equation
$$x - 7 = 9,$$
the number 7 being connected with the unknown quantity x by the sign —, the one is *subtracted* from the other. To reduce the equation by a contrary process, let 7 be *added* to both sides. It then becomes
$$x - 7 + 7 = 9 + 7.$$
The equality of the members is preserved, because one is as much increased as the other. (Axiom 1.) But on one side, we have —7 and +7. As these are equal, and have contrary signs, they *balance each other*, and may be cancelled. (Art. 77.) The equation will then be
$$x = 9 + 7.$$
Here the value of x is found. It is shown to be equal to $9+7$, that is to 16. The equation is therefore reduced. The unknown quantity is on one side by itself, and all the known quantities on the other side.

In the same manner if $x - b = a$
Adding b to both sides $x - b + b = a + b$
And cancelling $(-b+b)$ $x = a + b.$

Here it will be seen that the last equation is the same as

the first, except that b is on the opposite side, with a contrary sign.

Next suppose

$$y + c = d$$

Here c is *added* to the unknown quantity y. To reduce the equation by a contrary process, let c be *subtracted* from both sides, that is, let $-c$ be applied to both sides. We then have

$$y + c - c = d - c$$

The equality of the members is not affected, because one is as much diminished as the other. (Ax. 2.) When $(+c-c)$ is cancelled, the equation is reduced, and is

$$y = d - c.$$

This is the same as $y + c = d$, except that c has been transposed, and has received a contrary sign. We hence obtain the following general rule :

173. *When known quantities are connected with the unknown quantity by the sign $+$ or $-$, the equation is reduced by* TRANSPOSING *the known quantities to the other side, and prefixing the contrary sign.*

This is called reducing an equation by *addition* or *subtraction*, because it is, in effect, adding or subtracting certain quantities, to or from, each of the members.

Ex. 1. Reduce the equation $\qquad x + 3b - m = h - d$
　　　　Transposing $+3b$, we have : $\quad x - m = h - d - 3b$
　　　　And transposing $-m$, $\qquad\quad x = h - d - 3b + m$

174. When several terms on the same side of an equation are *alike*, they may be united in one, by the rules for reduction in addition. (Art. 72 and 74.)

Ex. 2. Reduce the equation $\qquad x + 5b - 4h = 7b$
　　　　Transposing $5b - 4h$ $\qquad\quad x = 7b - 5b + 4h$
　　　　Uniting $7b - 5b$ in one term $\quad x = 2b + 4h$

175. The *unknown* quantity must also be transposed, whenever it is on both sides of the equation. It is not material on which side it is finally placed. For if $x = 3$; it is evident that $3 = x$. It may be well however, to bring it on that side, where it will have the the affirmative sign, when the equation is reduced.

Ex. 3. Reduce the equation $\qquad 2x+2h=h+d+3x$
By transposition $\qquad 2h-h-d=3x-2x$
Uniting terms $\qquad h-d=x.$

176. When the *same term* is on *opposite sides* of the equation, instead of transposing, we may *expunge* it from each. For this is only subtracting the same quantity from equal quantities. (Ax. 2.)

Ex. 4. Reduce the equation $\qquad x+3h+d=b+3h+7d$
Expunging $3h$ $\qquad x+d=b+7d$
Transp. and uniting terms $\qquad x=b+6d.$

177. As *all* the terms of an equation may be transposed, or supposed to be transposed; and it is immaterial which member is written first; it is evident that the *signs of all the terms may be changed*, without affecting the equality.

Thus, if we have $\qquad x-b=d-a$
Then by transposition $\qquad -d+a=-x+b$
Or, inverting the members $\qquad -x+b=-d+a.$

178. If all the terms on *one* side of an equation be transposed, each member will be equal to 0.

Thus if $x+b=d,$ $\qquad$ then $x+b-d=0.$

It is frequently convenient to reduce an equation to this form, in which the positive and negative terms *balance* each other. In the example just given, $x+b$ is balanced by $-d.$ For in the first of the two equations, $x+b$ is *equal* to $d.$

Ex. 5. Reduce $a+2x-8=b-4+x+a.$

6. Reduce $y+ab-hm=a+2y-ab+hm.$

7. Reduce $h+30+7x=8-6h+6x-d+b.$

8. Reduce $bh-21-4x+d=12-3x+d+7bh.$

REDUCTION OF EQUATIONS BY MULTIPLICATION.

179. The unknown quantity, instead of being connected with a known quantity by the sign $+$ or $-$, may be *divided* by it, as in the equation

$$\frac{x}{a}=b.$$

L

Here the reduction can not be made, as in the preceding instances, by transposition. But if both members be *multiplied* by *a*, (Art. 170.) the equation will become

$$x = ab.$$

For *a fraction is multiplied into its denominator, by removing the denominator.* This has been proved from the properties of fractions. (Art. 159.) It is also evident from the sixth axiom.

Thus $x = \dfrac{ax}{a} = \dfrac{3x}{3} = \dfrac{(a+b) \times x}{a+b} = \dfrac{dx+5x}{d+5}$, &c. For in each of these instances, x is both multiplied and divided by the same quantity; and this makes no alteration in the value. Hence,

180. *When the unknown quantity is* DIVIDED *by a known quantity, the equation is reduced by* MULTIPLYING *each side by this known quantity.*

The same transpositions are to be made in this case, as in the preceding examples. It must be observed also, that *every* term of the equation is to be multiplied. For the several terms in each member constitute a compound multiplicand, which is to be multiplied according to art. 98.

Ex. 1. Reduce the equation, $\qquad \dfrac{x}{c} + a = b + d$

Multiplying both sides by $\qquad c$.

The product is $\qquad\qquad x + ac = bc + cd$
And transposing ac $\qquad\quad x = bc + cd - ac.$

2. Reduce the equation $\qquad \dfrac{x-4}{6} + 5 = 20$

Multiplying by 6 $\qquad\qquad x - 4 + 30 = 120$
Transp. and uniting terms $\quad x = 120 + 4 - 30 = 94.$

3. Reduce the equation $\qquad \dfrac{x}{a+b} + d = h$

Multiplying by $a+b$ (Art. 100.) $x + ad + bd = ah + bh$
By transposition $\qquad\qquad x = ah + bh - ad - bd.$

181. When the unknown quantity is in the *denominator* of a fraction, the reduction is made in a similar manner, by multiplying the equation by this denominator.

Ex. 4. Reduce the equation $\dfrac{6}{10-x}+7=8$

Multiplying by $10-x$ $6+70-7x=80-8x$

Transp. and uniting terms $x=4.$

182. Though it is not generally *necessary*, yet it is often convenient, to remove the denominator from a fraction consisting of *known* quantities only. This may be done, in the same manner, as the denominator is removed from a fraction which contains the unknown quantity.

Take for example $\dfrac{x}{a}=\dfrac{d}{b}+\dfrac{h}{c}$

Multiplying by a (Art. 158.) $x=\dfrac{ad}{b}+\dfrac{ah}{c}$

Multiplying by b $bx=ad+\dfrac{abh}{c}$

Multiplying by c $bcx=acd+abh.$

Or we may multiply by the product of *all* the denominators at once.

In the same equation $\dfrac{x}{a}=\dfrac{d}{b}+\dfrac{h}{c}$

Multiplying by abc $\dfrac{abcx}{a}=\dfrac{abcd}{b}+\dfrac{abch}{c}$

Then by cancelling from each term, the letter which is common to its numerator and denominator, (Art. 145,) we have $bcx=acd+abh$, as before. Hence,

183. *An equation may be cleared of* FRACTIONS *by multiplying each side into all the* DENOMINATORS.

Thus the equation $\dfrac{x}{a}=\dfrac{b}{d}+\dfrac{e}{g}-\dfrac{h}{m}$

Is the same as $dgmx=abgm+adem-adgh.$

And the equation $\dfrac{x}{2}=\dfrac{2}{3}+\dfrac{4}{5}+\dfrac{6}{2}$

Is the same as $30x=40+48+180.$

REDUCTION OF EQUATIONS BY DIVISION.

184. *When the unknown quantity is* MULTIPLIED *into any known quantity, the equation is reduced by* DIVIDING *both sides by this known quantity.* (Ax. 4.)

Ex. 1. Reduce the equation $\quad\quad ax+b-3h=d$

By transposition $\quad\quad\quad ax=d+3h-b$

Dividing by a $\quad\quad\quad\quad\quad\quad x=\dfrac{d+3h-b}{a},$

2. Reduce the equation $\quad\quad 2x=\dfrac{a}{c}-\dfrac{d}{h}+4b$

Clearing of fractions (Art. 183.) $2chx=ah-cd+4bch$

Dividing by $2ch$ $\quad\quad\quad\quad x=\dfrac{ah-cd+4bch}{2ch}$

185. If the unknown quantity has co-efficients in *several* terms, the equation must be divided by *all* these co-efficients, connected by their signs, according to art. 121.

Ex. 3. Reduce the equation $\quad\quad 3x+d=bx+a$

By transposition $\quad\quad\quad\quad 3x-bx=a-d$

That is, (Art. 120.) $\quad\quad (3-b)\times x=a-d$

Dividing by $3-b$ $\quad\quad\quad\quad x=\dfrac{a-d}{3-b}$

Ex. 4. Reduce the equation $\quad\quad ax+4=h-x$

By transposition $\quad\quad\quad\quad ax+x=h-4$

That is $\quad\quad\quad\quad\quad (a+1)\times x=h-4$

Dividing by $a+1$ $\quad\quad\quad\quad x=\dfrac{h-4}{a+1}$

Ex. 5. Reduce the equation $\quad\quad x-\dfrac{x-b}{h}=\dfrac{a+d}{4}$

Clearing of fractions. See art. 142. $\quad 4hx-(4x-4b)=ah+dh$

That is (Art. 82.) $\quad\quad\quad\quad 4hx-4x+4b=ah+dh$

Transposing $4b$ $\quad\quad\quad\quad 4hx-4x=ah+dh-4b$

Dividing by $4h-4$ $\quad\quad\quad\quad x=\dfrac{ah+dh-4b}{4h-4}$

186. If any quantity, either known or unknown, is found as a factor in *every term*, the equation may be *divided* by it. On the other hand, if any quantity is a *divisor* in every term, the equation may be *multiplied* by it. In this way, the factor or divisor will be removed, so as to render the expression more simple,

Ex. 6. Reduce the equation $\qquad ax+3ab=6ad+a$
Dividing by a (Arts. 120 and 128.) $\quad x+3b=6d+1$
And transposing $3b$ $\qquad\qquad\qquad x=6d+1-3b$

7. Reduce the equation $\qquad\qquad \dfrac{x+1}{x}-\dfrac{b}{x}=\dfrac{h-d}{x}$

Multiplying by x (Art. 159.) $\qquad x+1-b=h-d$
And transposing $1-b$ $\qquad\qquad x=h-d+b-1.$

8. Reduce the equation $\qquad x\times(a+b)-a+b=d\times(a+b)$
Dividing by $a+b$ (Art. 118.) $\quad x-1=d$
Transposing -1 $\qquad\qquad\quad x=d+1$

---

187. Sometimes the conditions of a problem are at first stated, not in an equation, but by means of a *proportion.* To show how this may be reduced to an equation, it will be necessary to anticipate the subject of a future section, so far as to admit the principle, that "when four quantities are in geometrical proportion, the product of the two extremes is equal to the product of the two means:" a principle which is at the foundation of the Rule of Three in arithmetic. See Webber's Arithmetic.

Thus if $a:b::c:d$, $\qquad\qquad$ Then $ad=bc$
And if $3:4::6:8$; $\qquad\qquad$ And $3\times8=4\times6$. Hence,

188. *A proportion is converted into an equation, by making the product of the extremes, one side of the equation; and the product of the means, the other side.*

Ex. 1. Reduce to an equation $\qquad\qquad ax:b::ch:d.$
The product of the extremes is $\qquad adx$
The product of the means is $\qquad bch$
The equation is, therefore $\qquad adx=bch.$

2. Reduce to an equation $\qquad\qquad a+b:c::h-m:y.$
The product of the extremes is $\qquad ay+by$
The product of the means is $\qquad ch-cm$
The equation is, therefore $\qquad ay+by=ch-cm.$

189. On the other hand, *an equation may be converted into a proportion, by resolving one side of the equation into two factors, for the middle terms of the proportion; and the other side into two factors, for the extremes.*

As a quantity may often be resolved into different pairs of factors; (Art. 42.) a variety of proportions may frequently be derived from the same equation.

Ex. 1. Reduce to a proportion $abc = deh$

The side abc may be resolved into $a \times bc$, or $ab \times c$, or $ac \times b$. And deh may be resolved into $d \times eh$, or $de \times h$ or $dh \times e$.

Therefore $a:d::eh:bc$ And $ac:dh::e:b$

And $ab:de::h:c$ And $ac:d::eh:b$ &c.

For in each of these instances, the product of the extremes is abc, and the product of the means deh.

2. Reduce to a proportion $ax + bx = cd - ch$

The first member may be resolved into $x \times (a+b)$

And the second into $c \times (d-h)$

Therefore $x:c::d-h:a+b$ And $d-h:x::a+b:c$, &c.

190. If for any term or terms in an equation, any other expression of the same value be *substituted*, it is manifest that the equality of the sides will not be affected.

Thus, instead of 16, we may write 2×8 or $\dfrac{64}{4}$, or $25 - 9$, &c.

For these are only different forms of expression for the same quantity.

191. It will generally be well to have the several steps, in the reduction of equations, succeed each other in the following order.

First, Clear the equation of fractions. (Art. 183.)

Secondly, Transpose and unite the terms. (Arts. 173, 4, 5.)

Thirdly, Divide by the co-efficients of the unknown quantity. (Arts. 184, 5.)

EXAMPLES.

1. Reduce the equation $\dfrac{3x}{4} + 6 = \dfrac{5x}{8} + 7$

Clearing of fractions $24x + 192 = 20x + 224$

Transp. and uniting terms $4x = 32$

Dividing by 4 $x = 8$.

2. Reduce the equation $\dfrac{x}{a}+h=\dfrac{x}{b}-\dfrac{x}{c}+d$

Clearing of fractions . $bcx+abch=acx-abx+abcd$

By transposition . $bcx+abx-acx=abcd-abch$

Dividing by the co-eff's of x, $x=\dfrac{abcd-abch}{bc+ab-ac}.$

3. Reduce $40-6x-16=120-14x.$ Ans. $x=12.$

4. Reduce $\dfrac{x-3}{2}+\dfrac{x}{3}=20-\dfrac{x-19}{2}$ Ans. $x=\dfrac{93}{4}.$

5. Reduce $\dfrac{x}{3}+\dfrac{x}{5}=20-\dfrac{x}{4}.$ 6. Reduce $\dfrac{1-a}{x}-4=5.$

7. Reduce $\dfrac{3}{x+4}-2=8.$ 8. Reduce $\dfrac{6x}{x+4}=1.$

Solution of Problems.

192. In the solution of problems, by means of equations, two things are necessary : First to translate the statement of the question from common to algebraic language, in such a manner as to form an equation : Secondly, to reduce this equation to a state in which the unknown quantity will stand by itself, and its value be given in known terms, on the opposite side. The manner in which the latter is effected, has already been considered. The former will probably occasion more perplexity to a beginner ; because the conditions of questions are so various in their nature, that the proper method of stating them can not be easily learned, like the reduction of equations, by a system of definite rules. Practice however will soon remove a great part of the difficulty.

193. It is one of the principal peculiarities of an algebraic solution, that the *quantity sought* is itself introduced into the operation. This enables us to make a statement of the conditions, in the same form, as though the problem were already solved. Nothing then remains to be done, but to *reduce* the equation, and to find the aggregate value of the known quantities. (Art. 53.) As these are equal to the *unknown* quantity on the other side of the equation, the value

of that also is determined, and therefore the problem is solved.

Problem 1. A man being asked how much he gave for his watch, replied ; If you multiply the price by 4, and to the product add 70, and from this sum subtract 50, the remainder will be equal to 220 dollars.

To solve this, we must first translate the conditions of the problem, into such algebraic expressions, as will form an equation.

Let the price of the watch be represented by x

This price is to be mult'd by 4, which makes $4x$

To the product, 70 is to be added, making $4x+70$

From this, 50 is to be subtracted, making $4x+70-50$

Here we have a number of the conditions, expressed in algebraic terms; but have as yet no *equation*. We must observe then, that by the last condition of the problem, the preceding terms are said to be *equal to* 220.

We have, therefore, this equation $4x+70-50=220$.
To reduce this,

Transpose and unite the terms, then $4x=200$

Divide by 4 (Art. 184.) and $x=50$.

Here the value of x is found to be 50 dollars, which is the price of the watch.

194. To *prove* whether we have obtained the true value of the letter which represents the unknown quantity, we have only to substitute this value, for the letter itself, in the equation which contains the first statement of the conditions of the problem; and to see whether the sides are equal, after the substitution is made. For if the answer thus satisfies the conditions proposed, it is the quantity sought. Thus, in the preceding example,

The original equation is $4x+70-50=220$

Substituting 50 for x, it becomes $4\times50+70-50=220$.

To see whether the first member of this equation is equal to the second,

Multiply 4 into 50 ; the product is	200
To this add	70
The sum is	270
From this subtract	50
The remainder is	220 as in

the statement of the problem.

Prob. 2. What number is that, to which if its half be added, and from the sum 20 be subtracted, the remainder will be a fourth part of the number itself?

In stating questions of this kind, where fractions are concerned, it should be recollected, that $\frac{1}{3}x$ is the same as $\frac{x}{3}$; that $\frac{2}{5}x=\frac{2x}{5}$, &c. (Art. 161.)

In this problem, let x be put for the number required.

Then by the conditions proposed, $\qquad x+\dfrac{x}{2}-20=\dfrac{x}{4}$

Clearing of fractions $\qquad\qquad\qquad 8x+4x-160=2x$

Transp. and uniting terms $\qquad\qquad 10x=160$

Dividing by 10 $\qquad\qquad\qquad\qquad x=16.$

$\qquad\qquad$ Proof $\qquad\qquad\qquad 16+\dfrac{16}{2}-20=\dfrac{16}{4}.$

Prob. 3. A father divides his estate among his three sons, in such a manner, that;

The first has $1000 less than half of the whole;

The second has 800 less than a third of the whole;

The third has 600 less than a fourth of the whole;

What is the value of the estate?

If the whole estate be represented by x, then the several shares will be $\dfrac{x}{2}-100$, and $\dfrac{x}{3}-800$, and $\dfrac{x}{4}-600.$

And as these constitute the whole estate, they are together equal to x.

We have then this equation $\dfrac{x}{2}-1000+\dfrac{x}{3}-800+\dfrac{x}{4}-600=x$

Or, by uniting terms, $\qquad \dfrac{x}{2}+\dfrac{x}{3}+\dfrac{x}{4}-2400=x$

Clearing of fractions $\qquad 12x+8x+6x-57600=24x$

Transp. and uniting terms $\quad 2x=57600$

Dividing by 2 $\qquad\qquad\qquad x=28800$

Proof. $\dfrac{28800}{2}-1000+\dfrac{28800}{3}-800+\dfrac{28800}{4}-600=28800.$

195. To avoid an unnecessary introduction of unknown quantities into an equation, it may be well to observe, in this place, that when the *sum* or *difference* of two quantities is

M

given, both of them may be expressed by means of the same letter. For if one of two quantities be subtracted from their sum, it is evident the remainder will be equal to the other. And if the difference of two quantities be subtracted from the greater, the remainder will be the less.

Thus if the sum of two numbers be   20
And if one of them be represented by x
The other will be equal to $20-x$.

Prob. 4. Divide 48 into two such parts, that if the less be divided by 4 and the greater by 6, the sum of the quotients will be 9.

Here if x be put for the smaller part, the greater will be $48-x$.

By the conditions of the problem $\dfrac{x}{4}+\dfrac{48-x}{6}=9$.

Clearing of fractions $6x+192-4x=216$
Transposing and uniting terms $2x=24$
Dividing by 2 $x=12$, the less.
Then $48-x=48-12=36$, the greater.

196. Letters may be employed to express the *known* quantities in an equation, as well as the unknown. A particular value is assigned to the numbers, when they are introduced into the calculation: and at the close, the numbers are restored. (Art. 52.)

Prob. 5. If, to a certain number, 720 be added, and the sum be divided by 125; the quotient will be equal to 7392 divided by 462. What is that number?

Let $x=$ the number required.

$a=720$ $d=7392$
$b=125$ $h=462$

Then by the conditions of the problem $\dfrac{x+a}{b}=\dfrac{d}{h}$.

Clearing of fractions $hx+ah=bd$
Transposing ah $hx=bd-ah$

Dividing by h $x=\dfrac{bd-ah}{h}$

Restoring the numbers, $x=\dfrac{(125\times7392)-(720\times462)}{462}=1280$.

197. When the resolution of an equation brings out a *negative* answer, it shows that the value of the unknown quantity is *contrary* to the quantities which, in the statement of the question, are considered positive. See Negative Quantities. (Art. 54, &c.)

Prob. 6. A merchant gains or loses, in a bargain, a certain sum. In a second bargain, he gains 350 dollars, and, in a third, loses 60. In the end, he finds he has gained 200 dollars, by the three together. How much did he gain or lose by the first?

In this example, as the profit and loss are opposite in their nature, they must be distinguished by contrary signs. (Art. 57.) If the profit is marked +, the loss must be —.

Let $x =$ the sum required.

Then according to the statement	$x + 350 - 60 = 200$
By transposition	$x = 200 + 60 - 350$
And uniting the terms	$x = -90.$

The negative sign prefixed to the answer, shows that there was a *loss* in the first bargain; and therefore that the proper sign of x is negative also. But this being determined by the answer, the omission of it in the course of the calculation can lead to no mistake.

Prob. 7. A ship sails 4 degrees north, then 13 S. then 17 N. then 19 S. and has finally 11 degrees of south latitude. What was her latitude at starting?

Let $x =$ the latitude sought.

Then marking the northings +, and the southings — ;

By the statement	$x + 4 - 13 + 17 - 19 = -11$
By transposition	$x = 13 + 19 - 11 - 4 - 17$
And uniting terms	$x = 0.$

The answer here shows that the place from which the ship started was on the equator, where the latitude is nothing.

Prob. 8. If a certain number is divided by 12, the quotient, dividend, and divisor added together will amount to 64. What is the number?

Let $x =$ the number sought.

Then $\dfrac{x}{12}+x+12=64$

Multiplying by 12, (Art. 180.) $x+12x+144=768$
Transposing and uniting terms $13x=624$

Dividing by 13 $x=\dfrac{624}{13}=48.$

Prob. 9. An estate is divided among four children, in such a manner, that
The first has 200 dollars more than $\frac{1}{4}$ of the whole,
The second has 340 dollars more than $\frac{1}{5}$ of the whole,
The third has 300 dollars more than $\frac{1}{6}$ of the whole,
The fourth has 400 dollars more than $\frac{1}{8}$ of the whole.
What is the value of the estate?   Ans. 4800 dollars.

Prob. 10. What is that number which is as much less than 500, as a fifth part of it is greater than 40? Ans. 450.

Prob. 11. There are two numbers whose difference is 40, and which are to each other as 6 to 5. What are the numbers?
Let $x=$ the greater.
Then $x-40=$ the less. (Art. 195.)

By the conditions of the question $x:x-40::6:5$
Mult'g extremes and means (Art. 188.) $6x-240=5x$
Transposing and uniting terms $x=240,$ the greater.
And $240-40=200,$ the less.

Prob. 12. Three persons, A, B, and C draw prizes in a lottery. A draws 200 dollars; B draws as much as A, together with a third of what C draws; and C draws as much as A and B both. What is the amount of the three prizes.
Ans. 1200 dollars.

Prob. 13. What number is that, which is to 12 increased by three times the number, as 2 to 9? Ans. 8.

Prob. 14. A ship and a boat are descending a river at the same time. The ship passes a certain fort, when the boat is 13 miles below. The ship descends five miles, while the boat descends three. At what distance below the fort, will they be together?

Let $x=$ the distance required.

Then by the question	$x:x-13::5:3$
Mult. extremes and means	$5x-65=3x$
Transp. and uniting terms	$2x=65$
Dividing by 2	$x=32\frac{1}{2}.$

Prob. 15. What number is that, a sixth part of which exceeds an eighth part of it by 20? Ans. 480.

Prob. 16. Divide a prize of 2000 dollars into two such parts, that one of them shall be to the other, as 9:7.
Ans. The parts are 1125, and 875.

Prob. 17. What sum of money is that, whose third part, fourth part, and fifth part, added together, amount to 94 dollars? Ans. 120 dollars.

Prob. 18. Two travellers, A and B, 360 miles apart, travel towards each other till they meet. A's progress is 10 miles an hour, and B's 8. How far does each travel before they meet? Ans. A goes 200 miles, and B 160.

Prob. 19. A man spent one third of his life in England, one fourth of it in Scotland, and the remainder of it, which was 20 years, in the United States. To what age did he live? Ans. To the age of 48.

For the solution of many algebraic problems, an acquaintance with the calculations of powers and radical quantities is required. It will therefore be necessary to attend to these, before finishing the subject of equations.

INVOLUTION and POWERS.

ART. 198. WHEN *a quantity is multiplied into* ITSELF, *the* PRODUCT *is called a* POWER.

Thus $2 \times 2 = 4$, the square or second power of 2.
$2 \times 2 \times 2 = 8$, the cube or third power.
$2 \times 2 \times 2 \times 2 = 16$, the fourth power, &c.

So $10 \times 10 = 100$, the second power of 10.
$10 \times 10 \times 10 = 1000$, the third power.
$10 \times 10 \times 10 \times 10 = 10000$, the fourth power, &c.

And $a \times a = aa$, the second power of a.
$a \times a \times a = aaa$, the third power.
$a \times a \times a \times a = aaaa$, the fourth power, &c.

199. The original quantity itself, though not, like the powers proceeding from it, produced by multiplication, is nevertheless called the *first power*. It is also called the *root* of the other powers, because it is that from which they are all derived.

200. As it is inconvenient, especially in the case of high powers to write down all the letters or factors of which the powers are composed, an abridged method of notation is generally adopted. The root is written only once; and then a number or letter is placed at the right hand, and a little elevated, to signify how many times the root is *employed as a factor*, to produce the power. This number or letter is called the *index* or *exponent* of the power. Thus a^2 is put for $a \times a$ or aa, because the root a is *twice* repeated as a factor, to produce the power aa. And a^3 stands for aaa; for here a is repeated *three times* as a factor.

The index of the *first* power is 1; but this is commonly omitted. Thus a^1 is the same as a.

201. Exponents must not be confounded with *co-efficients*.

A co-efficient shows how often a quantity is taken as a *part* of a whole. An exponent shows how often a quantity is taken as a *factor* in a product.

Thus $4a=a+a+a+a$. But $a^4=a\times a\times a\times a$.

202. The scheme of notation by exponents has the peculiar advantage of enabling us to express an *unknown* power. For this purpose the index is a *letter*, instead of a numeral figure. In the solution of a problem, a quantity may occur, which we know to be *some* power of another quantity. But it may not be yet ascertained whether it is a square, a cube, or some higher power. Thus in the expression a^x, the index x denotes that a is involved to *some* power, though it does not determine *what* power. So b^m, and d^n are powers of b and d; and are read the mth power of b, and the nth power of d. When the value of the index is found, a *number* is generally substituted for the letter. Thus if $m=3$, then $b^m=b^3$: but if $m=5$, then $b^m=b^5$.

203. The method of expressing powers by exponents is also of great advantage in the case of *compound* quantities. Thus

$\overline{a+b+d}\,|^3$, or $\overline{a+b+d}^3$ or $(a+b+d)^3$, is $(a+b+d)\times(a+b+d) \times(a+b+d)$ that is, the cube of $(a+b+d)$. But this involved at length would be $[d^3$

$a^3+3a^2b+3a^2d+3ab^2+6abd+3ad^2+b^3+3b^2d+3bd^2+$

204. If we take a series* of powers whose indices increase or decrease by 1, we shall find that the powers themselves increase by a *common multiplier*, or decrease by a *common divisor*; and that this multiplier or divisor is the original quantity from which the powers are raised.

Thus in the series $aaaaa$, $aaaa$, aaa, aa, a;
Or $\qquad\qquad\qquad a^5\qquad a^4\qquad a^3\qquad a^2\qquad a^1$;

the indices counted from right to left are 1, 2, 3, 4, 5; and the common difference between them is a unit. If we begin on the *right*, and *multiply* by a, we produce the several powers, in succession, from right to left.

Thus $a\times a=a^2$ the 2d term.   And $a^3\times a=a^4$.
$a^2\times a=a^3$ the 3d term. $a^4\times a=a^5$, &c.

If we begin on the *left*, and *divide* by a,

*Note. The term *series* is applied to a number of quantities succeeding each other, in some regular order. It is not confined to any particular law of increase or decrease.

We have $a^5 \div a = a^4$.　　　　And $a^3 \div a = a^2$.
　　　　$a^4 \div a = a^3$.　　　　　　$a^2 \div a = a^1$.

205. But this division may be carried still farther; and we shall then obtain a new set of quantities.

Thus $a \div a = \dfrac{a}{a} = 1$. (Art. 128.) And $\dfrac{1}{a} \div a = \dfrac{1}{aa}$. (Art.163.)

$1 \div a = \dfrac{1}{a}$.　　　　　　$\dfrac{1}{aa} \div a = \dfrac{1}{aaa}$, &c.

The whole series then

Is $aaaaa,\ aaaa,\ aaa,\ aa,\ a,\ 1, \dfrac{1}{a},\ \dfrac{1}{aa},\ \dfrac{1}{aaa}$, &c.

Or $a^5,\quad a^4,\quad a^3,\quad a^2,\ a,\ 1, \dfrac{1}{a},\ \dfrac{1}{a^2},\ \dfrac{1}{a^3}$, &c.,

Here the quantities on the *right* of 1, are the *reciprocals* of those on the *left*. (Art. 49.) The former, therefore, may be properly called *reciprocal powers* of a; while the latter may be termed, for distinction sake, *direct powers* of a. It may be added, that the powers on the left are also the reciprocals of those on the right.

For $1 \div \dfrac{1}{a} = 1 \times \dfrac{a}{1} = a$. (Art. 162.)　And $1 \div \dfrac{1}{a^3} = a^3$.

$1 \div \dfrac{1}{a^2} = 1 \times \dfrac{a^2}{1} = a^2$.　　　　$1 \div \dfrac{1}{a^4} = a^4$, &c.

206. The same plan of notation is applicable to *compound* quantities. Thus, from $a+b$, we have the series,

$(a+b)^3,\ (a+b)^2,\ (a+b),\ 1, \dfrac{1}{a+b},\ \dfrac{1}{(a+b)^2},\ \dfrac{1}{(a+b)^3}$, &c

207. For the convenience of calculation, another form of notation is given to reciprocal powers.

According to this, $\dfrac{1}{a}$ or $\dfrac{1}{a^1} = a^{-1}$.　And $\dfrac{1}{aaa}$ or $\dfrac{1}{a^3} = a^{-3}$.

$\dfrac{1}{aa}$ or $\dfrac{1}{a^2} = a^{-2}$.　　　　$\dfrac{1}{aaaa}$ or $\dfrac{1}{a^4} = a^{-4}$. &c.

And to make the indices a complete series, with 1 for the

common difference, the term $\dfrac{a}{a}$ or 1, which is considered as *no* power, is written a^0.

The powers both direct and reciprocal* then,

Instead of $aaaa,\ aaa,\ aa,\ a,\ \dfrac{a}{a},\ \dfrac{1}{a},\ \dfrac{1}{aa},\ \dfrac{1}{aaa},\ \dfrac{1}{aaaa},$ &c.

Will be $\quad a^4,\ a^3\ a^2, a^1, a^0, a^{-1}, a^{-2}, a^{-3},\quad a^{-4}$, &c.

Or $\quad\ a^{+4}, a^{+3}, a^{+2}, a^{+1}, a^0, a^{-1}, a^{-2}, a^{-3},\quad a^{-4}$, &c.

And the indices taken by themselves will be,

$$+4,\ +3,\ +2,\ +1,\ 0,\ -1,\ -2,\ -3,\ -4,\ \&c.$$

208. The root of a power may be expressed by more letters than one.

Thus $aa \times aa$, or $\overline{aa}\,|^2$ is the second power of aa.

And $aa \times aa \times aa$, or $\overline{aa}\,|^3$ is the third power of aa, &c.

Hence a certain power of one quantity, may be a different power of another quantity. Thus a^4 is the second power of a^2, and the fourth power of a.

209. All the powers of 1 are the same. For 1×1, or $1 \times 1 \times 1$, &c. is still 1.

INVOLUTION.

210. Involution is finding any power of a quantity, by multiplying it into itself. The reason of the following general rule is manifest, from the nature of powers.

Multiply the quantity into itself, till it is taken as a factor, as many times as there are units in the index of the power to which the quantity is to be raised.

This rule comprehends all the instances which can occur in involution. But it will be proper to give an explanation of the manner in which it is applied to particular cases.

211. A single letter is involved, by giving it the index of the proposed power; or by repeating it as many times, as there are units in that index.

The 4th power of a, is a^4 or $aaaa$. (Art. 198.)

The 6th power of y, is y^6 or $yyyyyy$.

The nth power of x, is x^n or $xxx \dots n$ times repeated.

* See note B.

N

212. The method of involving a quantity which consists of several *factors*, depends on the principle, that *the power of the product of several factors is equal to the product of their powers.*

Thus $(ay)^2 = a^2 y^2$. For by art. 210; $(ay)^2 = ay \times ay$.

But $ay \times ay = ayay = aayy = a^2 y^2$.

So $(bmx)^3 = bmx \times bmx \times bmx = bbbmmmxxx = b^3 m^3 x^3$.

And $(ady)^n = ady \times ady \times ady \dots n$ times $= a^n d^n y^n$.

In finding the power of a product, therefore, we may either involve the whole at once; or we may involve each of the factors separately, and then multiply their several powers into each other.

Ex. 1. The 4th power of dhy, is $(dhy)^4$, or $d^4 h^4 y^4$.

2. The 3d power of $4b$, is $(4b)^3$, or $4^3 b^3$, or $64b^3$.

3. The nth power of $6ad$, is $(6ad)^n$, or $6^n a^n d^n$.

4. The 3d power of $3m \times 2y$, is $(3m \times 2y)^3$, or $27m^3 \times 8y^3$.

213. A compound quantity, consisting of terms connected by $+$ and $-$, is involved by an actual multiplication of its several parts. Thus,

$(a+b)^1 = a+b$, the first power.

$$\begin{array}{l} a^2 + \ ab \\ \ \ \ + \ ab + b^2 \\ \hline \end{array}$$

$(a+b)^2 = a^2 + 2ab + b^2$, the second power of $(a+b)$.

$a \ + b$

$$\begin{array}{l} a^3 + 2a^2 b + \ ab^2 \\ \ \ \ + \ a^2 b + 2ab^2 + b^3 \\ \hline \end{array}$$

$(a+b)^3 = a^3 + 3a^2 b + 3ab^2 + b^3$, the 3d power.

$a \ + \ b$

$$\begin{array}{l} a^4 + 3a^3 b + 3a^2 b^2 + ab^3 \\ \ \ \ + \ a^3 b + 3a^2 b^2 + 3ab^3 + b^4 \\ \hline \end{array}$$

$(a+b)^4 = a^4 + 4a^3 b + 6a^2 b^2 + 4ab^3 + b^4$; the 4th power, &c.

2. The square of $a-b$, is $a^2-2ab+b^2$.

3. The cube of $a+1$, is a^3+3a^2+3a+1.

4. The square $a+b+h$, is $a^2+2ab+2ah+b^2+2bh+h^2$.

5. Required the cube of $a+2d+3$.

6. Required the 4th power of $b+2$.

214. The squares of *binomial* and *residual* quantities occur so frequently in algebraic processes, that it is important to make them familiar.

If we multiply $a+h$ into itself, and also $a-h$,

We have $a+h$		And $a-h$
$a+h$		$a-h$

$$a^2+ah$$
$$+ah+h^2$$

$$a^2-ah$$
$$-ah+h^2$$

$$a^2+2ah+h^2.$$

$$a^2-2ah+h^2.$$

Here it will be seen that, in each case, the first and last terms are squares of a and h; and that the middle term is twice the product of a into h. Hence the squares of binomial and residual quantities, without multiplying each of the terms separately, may be found, by the following proposition.*

The square of a binomial, the terms of which are both positive, is equal to the square of the first term, + twice the product of the two terms; + the square of the last term.

And the square of a *residual* quantity, is equal to the square of the first term, — twice the product of the two terms, + the square of the last term.

Ex. 1. The square of $2a+b$, is $4a^2+4ab+b^2$.

2. The square of $h+1$, is h^2+2h+1.

3. The square of $ab+cd$, is $a^2b^2+2abcd+c^2d^2$.

4. The square of $6y+3$, is $36y^2+36y+9$.

5. The square of $3d-h$, is $9d^2-6dh+h^2$.

6. The square of $a-1$, is a^2-2a+1.

For the method of finding the higher powers of binomials, see one of the succeeding sections.

* Euclid's Elements, Book II. Prop. 4.

215. For many purposes, it will be sufficient to express the powers of compound quantities by *exponents*, without an actual multiplication.

Thus the square of $a+b$, is $\overline{a+b}|^2$, or $(a+b)^2$. Art. 203.

The nth power of $bc+8+x$, is $(bc+8+x)^n$.

In cases of this kind, the vinculum must be drawn over *all* the terms of which the compound quantity consists.

216. But if the root consists of several *factors*, the vinculum which is used in expressing the power, may either extend over the whole; or may be applied to each of the factors separately, as convenience may require.

Thus the square of $\overline{a+b}\times\overline{c+d}$, is either

$$\overline{\overline{a+b}\times\overline{c+d}}|^2 \text{ or } \overline{a+b}|^2 \times\overline{c+d}|^2.$$

For the first of these expressions is the square of the product of the two factors, and the last is the product of their squares. But one of these is equal to the other. (Art. 212.)

The cube of $a\times\overline{b+d}$, is $(a\times\overline{b+d})^3$, or $a^3\times(b+d)^3$.

217. When a quantity, whose power has been expressed by a vinculum and an index, is afterwards involved by an actual multiplication of the terms, it is said to be *expanded*.

Thus $(a+b)^2$, when expanded, becomes $a^2+2ab+b^2$.

And $(a+b+h)^2$, becomes $a^2+2ab+2ah+b^2+2bh+h^2$.

218. With respect to the *sign* which is to be prefixed to quantities involved, it is important to observe, that *when the root is positive, all its powers are positive also ; but when the root is negative, the* ODD *powers are negative, while the* EVEN *powers are positive.*

For the proof of this, see art. 109.

The 2d power of $-a$ is $+a^2$
The 3d power is $-a^3$
The 4th power is $+a^4$
The 5th power is $-a^5$, &c.

219. Hence any *odd* power has the same sign as its root. But an *even* power is positive, whether its root is positive or negative.

Thus $+a\times+a=a^2$.
And $-a\times-a=a^2$.

220. *A quantity which is already a power, is involved by multiplying its index, into the index of the power to which it is to be raised.*

1. The 3d power of a^2, is $a^{2\times3}=a^6$

For $a^2 = aa$; and the cube of aa is $aa \times aa \times aa = aaaaaa = a^6$; which is the 6th power of a, but the 3d power of a^2.

For a farther illustration of this rule, see arts. 233, 4.

2. The 4th power of $a^3 b^2$, is $a^{3 \times 4} b^{2 \times 4} = a^{12} b^8$.

3. The 3d power of $4a^2 x$, is $64 a^6 x^3$.

4. The 4th power of $2a^3 \times 3x^2 d$, is $16 a^{12} \times 81 x^8 d^4$,

5. The 5th power of $(a+b)^2$, is $(a+b)^{10}$.

6. The nth power of a^3, is a^{3n}.

7. The nth power of $(x-y)^m$, is $(x-y)^{mn}$.

8, $\overline{a^3 + b^3}|^2 = a^6 + 2a^3 b^3 + b^6$. (Art. 214.)

9. $\overline{a^3 \times b^3}|^2 = a^6 \times b^6$.   10. $(a^3 b^2 h^4)^3 = a^9 b^6 h^{12}$.

221. The rule is equally applicable to powers whose exponents are *negative*.

Ex. 1. The 3d power of a^{-2}, is $a^{-2 \times 3} = a^{-6}$.

For $a^{-2} = \dfrac{1}{aa}$. (Art. 207.)   And the 3d power of this is,

$$\frac{1}{aa} \times \frac{1}{aa} \times \frac{1}{aa} = \frac{1}{aaaaaa} = \frac{1}{a^6} = a^{-6}.$$

2. The 4th power of $a^2 b^{-3}$, is $a^8 b^{-12}$, or $\dfrac{a^8}{b^{12}}$.

3. The cube of $2x^n y^{-m}$, is $8x^{3n} y^{-3m}$.

4. The square of $b^3 x^{-1}$, is $b^6 x^{-2}$.

5. The nth power of x^{-m}, is x^{-mn}, or $\dfrac{1}{x^{mn}}$.

222. It must be observed here, as in art. 218, that if the sign which is *prefixed* to the power be −, it must be changed to +, whenever the index becomes an even number.

Ex. 1. The square $-a^3$, is $+a^6$. For the square of $-a^3$, is $-a^3 \times -a^3$, which, according to the rules for the signs in multiplication, is $+a^6$.

2. But the *cube* of $-a^3$, is $-a^9$. For $-a^3 \times -a^3 \times -a^3 = -a^9$.

3. The square of $-x^n$, is $+x^{2n}$.

4. The nth power of $-a^3$, is $\overset{+}{-} a^{3n}$.

Here the power will be positive or negative, according as the number which n represents is even or odd.

223. *A* FRACTION *is involved, by involving both the numerator, and the denominator.*

1. The square of $\dfrac{a}{b}$ is $\dfrac{a^2}{b^2}$. For, by the rule for the multiplication of fractions, (Art. 155.)

$$\frac{a}{b} \times \frac{a}{b} = \frac{aa}{bb} = \frac{a^2}{b^2}.$$

[(Art. 209.)

2. The 2d, 3d, and nth powers of $\dfrac{1}{a}$, are $\dfrac{1}{a^2}$, $\dfrac{1}{a^3}$ and $\dfrac{1}{a^n}$.

3. The cube of $\dfrac{2xr^2}{3y}$, is $\dfrac{8x^3r^6}{27y^3}$.

4. The nth power of $\dfrac{x^2 r}{ay^m}$, is $\dfrac{x^{2n}r^a}{a^n y^{mn}}$.

5. The square of $\dfrac{-a^3 \times (d+m)}{(x+1)^3}$, is $\dfrac{a^6 \times (d+m)^2}{(x+1)^6}$.

6. The cube of $\dfrac{-a^{-1}}{x^{-3}}$, is $\dfrac{-a^{-3}}{x^{-9}}$. (Art. 221.)

224. Examples of *binomials*, in which one of the terms is a fraction.

1. Find the square of $x+\frac{1}{2}$, and $x-\frac{1}{2}$, as in art. 214.

$$x+\tfrac{1}{2}$$
$$x+\tfrac{1}{2}$$

$$x^2 + \tfrac{1}{2}x$$
$$+ \tfrac{1}{2}x + \tfrac{1}{4}$$

$$x^2 + x + \tfrac{1}{4}.$$

$$x-\tfrac{1}{2}$$
$$x-\tfrac{1}{2}$$

$$x^2 - \tfrac{1}{2}x$$
$$- \tfrac{1}{2}x + \tfrac{1}{4}$$

$$x^2 - x + \tfrac{1}{4}.$$

2. The square of $a+\dfrac{2}{3}$, is $a^2 + \dfrac{4a}{3} + \dfrac{4}{9}$.

3. The square of $x+\dfrac{b}{2}$, is $x^2 + bx + \dfrac{b^2}{4}$.

4. The square of $x-\dfrac{b}{m}$, is $x^2 - \dfrac{2bx}{m} + \dfrac{b^2}{m^2}$.

225. It has been shown, (Art. 165.) that a *co-efficient* may be transferred, from the numerator to the denominator of a fraction, or from the denominator to the numerator. By re-

curring to the scheme of notation for reciprocal powers, (Art. 207.) it will be seen that *any factor* may also be transferred, *if the sign of its index be changed.*

1. Thus, in the fraction $\dfrac{ax^{-2}}{y}$, we may transfer x from the numerator to the denominator.

For $\dfrac{ax^{-2}}{y} = \dfrac{a}{y} \times x^{-2} = \dfrac{a}{y} \times \dfrac{1}{x^2} = \dfrac{a}{yx^2}.$

2. In the fraction $\dfrac{a}{by^3}$, we may transfer y from the denominator to the numerator.

For $\dfrac{a}{by^3} = \dfrac{a}{b} \times \dfrac{1}{y^3} = \dfrac{a}{b} \times y^{-3} = \dfrac{ay^{-3}}{b}.$

3. $\dfrac{da^{-4}}{x^3} = \dfrac{d}{x^3 a^4}.$

4. $\dfrac{b}{ay^2} = \dfrac{by^{-2}}{a}.$

226. In the same manner, we may transfer a factor which has a positive index in the numerator, or a negative index in the denominator.

1. Thus $\dfrac{ax^3}{b} = \dfrac{a}{bx^{-3}}.$ For x^3 is the reciprocal of x^{-3}, (Arts. 205, 207.) that is, $x^3 = \dfrac{1}{x^{-3}}.$ Therefore $\dfrac{ax^3}{b} = \dfrac{a}{bx^{-3}}.$

2. $\dfrac{h}{by^{-2}} = \dfrac{hy^2}{b}.$

3. $\dfrac{ad^2}{xy^{-3}} = \dfrac{ay^3}{xd^{-2}}.$

227. Hence, the denominator of any fraction may be entirely removed, or the numerator may be reduced to a unit, without altering the value of the expression.

1. Thus $\dfrac{a}{b} = \dfrac{1}{ba^{-1}},$ or $ab^{-1}.$

2. $\dfrac{x^{-2}}{b^{-n}} = \dfrac{1}{x^2 b^{-n}},$ or $b^n x^{-2}.$

3. $\dfrac{x^4 a^{-m}}{b^n c^{-3}} = \dfrac{1}{b^n c^m x^{-4} c^{-3}},$ or $c^3 x^4 a^{-m} b^{-n}.$

Addition and Subtraction of Powers.

228. It is obvious that powers may be added, like other quantitities, *by writing them one after another, with their signs.* (Art. 69.)

Thus the sum of a^3 and b^2, is $a^3 + b^2$.

And the sum of $a^2 - b^n$ and $h^5 - d^4$, is $a^2 - b^n + h^5 - d^4$.

229. The *same powers of the same letters* are *like quantities;* (Art. 45.) and their co-efficients may be added or subtracted, as in arts. 72 and 74.

Thus the sum of $2a^2$ and $3a^2$, is $5a^2$.

It is as evident that twice the square of a, and three times the square of a, are five times the square of a, as that twice a and three times a, are five times a.

To	$-3x^6y^5$	$3b^m$	$3a^4y^n$	$-5a^3h^6$	$3(a+y)^n$
Add	$-2x^6y^5$	$6b^m$	$-7a^4y^n$	$6a^3h^6$	$4(a+y)^n$
Sum	$-5x^6y^5$		$-4a^4y^n$		$7(a+y)^n$

230. But powers of *different letters,* and *different powers* of the *same letter,* must be added by writing them down with their signs.

The sum of a^3 and a^3, is $a^2 + a^3$.

It is evident that the square of a, and the cube of a, are neither twice the square of a, nor twice the cube of a.

The sum of $a^3 b^n$ and $3a^5 b^6$, is $a^3 b^n + 3a^5 b^6$.

231. *Subtraction* of powers is to be performed in the same manner as addition, except that the signs of the subtrahend are to be changed according to art. 82.

From	$2a^4$	$-3b^n$	$3h^2 b^6$	$a^3 b^n$	$5(a-h)^6$
Sub.	$-6a^4$	$4b^n$	$4h^2 b^6$	$a^n b^3$	$2(a-h)^6$
Diff.	$8a^4$		$-h^2 b^6$		$3(a-h)^6$

MULTIPLICATION OF POWERS.

232. Powers may be multiplied, like other quantities, by writing the factors one after another, either with, or without, the sign of multiplication between them. (Art. 93.)

Thus the product of a^3 into b^2, is $a^3 b^2$, or $aaabb$.

Mult.	x^{-3}	$h^2 b^{-a}$	$3a^6 y^2$	$dh^3 x^{-n}$	$a^2 b^3 y^2$
Into	a^m	a^4	$-2x$	$4by^4$	$a^3 b^2 y$
Prod.	$a^m x^{-3}$		$-6a^6 xy^2$		$a^2 b^3 y^2 a^3 b^2 y$

The product in the last example, may be abridged, by bringing together the letters which are repeated.

It will then become $a^5 b^5 y^3$.

The reason of this will be evident, by recurring to the series of powers in art. 207, viz.

$$a^{+4}, \quad a^{+3}, \quad a^{+2}, \quad a^{+1}, \quad a^0, \quad a^{-1}, \quad a^{-2}, \quad a^{-3}, \quad a^{-4}, \text{ \&c.}$$

Or, which is the same,

$$aaaa, \quad aaa, \quad aa, \quad a, \quad 1, \quad \frac{1}{a}, \quad \frac{1}{aa}, \quad \frac{1}{aaa}, \quad \frac{1}{aaaa}, \text{\&c.}$$

By comparing the several terms with each other, it will be seen that if any two or more of them be multiplied together, their product will be a power whose exponent is the *sum* of the exponents of the factors.

Thus $a^2 \times a^3 = aa \times aaa = aaaaa = a^5$.

Here 5, the exponent of the product, is equal to $2+3$, the sum of the exponents of the factors.

So $a^n \times a^m = a^{n+m}$.

For a^n, is a taken for a factor as many times, as there are units in n;

And a^m, is a taken for a factor as many times as there are units in m;

Therefore the product must be a, taken for a factor as many times, as there are units in both m and n. Hence,

233. *Powers of the same root may be multiplied, by adding their exponents.*

Q

Thus $a^2 \times a^6 = a^{2+6} = a^8$. And $x^3 \times x^2 \times x = x^{3+2+1} = x^6$.

Mult.	$4a^n$	$3x^4$	b^2y^3	$a^2b^3y^2$	$(b+h-y)^n$
Into	$2a^n$	$2x^3$	b^4y	a^3b^2y	$b+h-y$
Prod.	$8a^{2n}$		b^6y^4		$(b+h-y)^{n+1}$

234. The rule is equally applicable to powers whose ex-ponents are *negative*.

1. Thus $a^{-2} \times a^{-3} = a^{-5}$. That is $\dfrac{1}{aa} \times \dfrac{1}{aaa} = \dfrac{1}{aaaaa}$.

2. $y^{-n} \times y^{-m} = y^{-n-m}$. That is $\dfrac{1}{y^n} \times \dfrac{1}{y^m} = \dfrac{1}{y^n y^m}$.

3. $-a^{-2} \times a^{-3} = -a^{-5}$. That is $\dfrac{1}{-aa} \times \dfrac{1}{aaa} = \dfrac{1}{-aaaaa}$.

4. $a^{-2} \times a^3 = a^{3-2} = a^1$. That is $\dfrac{1}{aa} \times aaa = \dfrac{aaa}{aa} = a$.

In this example, the exponents are $+3$, and -2; and the sum of these is 1, according to the second case of reduction in addition. (Art. 74.)

5. $a^{-n} \times a^m = a^{m-n}$. That is $\dfrac{1}{a^n} \times a^m = \dfrac{a^m}{a^n}$.

6. $y^{-2} \times y^2 = y^0 = 1$. That is $\dfrac{1}{y^2} \times y^2 = \dfrac{y^2}{y^2} = 1$.

235. If $a+b$ be multiplied into $a-b$, the product will be $a^2 - b^2$: (Art. 110.) that is,

The product of the sum and difference of two quantities, is equal to the difference of their squares.

This is another instance of the facility with which *general truths* are demonstrated in algebra. See arts. 23 and 77.

If the sum and difference of the *squares* be multiplied, the product will be equal to the difference of the *fourth* powers, &c.

Thus $(a-y) \times (a+y) = a^2 - y^2$.
$(a^2 - y^2) \times (a^2 + y^2) = a^4 - y^4$.
$(a^4 - y^4) \times (a^4 + y^4) = a^8 - y^8$; &c.

DIVISION OF POWERS.

236. Powers may be divided, like other quantities, by re-jecting from the dividend a factor equal to the divisor; or by placing the divisor under the dividend, in the form of a fraction.

Thus the quotient of $a^3 b^2$ divided by b^2, is a^3. (Art.116.)

Divide	$9a^3 y^4$	$12b^3 x^n$	$a^2 b + 3a^2 y^4$	$d \times (a-h+y)^3$
By	$-3a^3$	$2b^3$	a^2	$(a-h+y)^3$
Quot.	$-3y^4$		$b+3y^4$	d

The quotient of a^5 divided by a^3, is $\dfrac{a^5}{a^3}$. But this is equal to a^2. For, in the series

$$a^{+4}, a^{+3}, a^{+2}, a^{+1}, a^0, a^{-1}, a^{-2}, a^{-3}, a^{-4}, \text{\&c.}$$

if any term be divided by another, the index of the quo-tient will be equal to the *difference* between the index of the dividend, and that of the divisor.

Thus $a^5 \div a^3 = \dfrac{aaaaa}{aaa} = aa = a^2$,

So $a^m \div a^n = \dfrac{a^m}{a^n} = a^{m-n}$. Hence,

237. A power may be divided by another power of the same root, by subtracting the index of the divisor from that of the dividend

Thus $y^3 \div y^2 = y^{3-2} = y^1$. That is $\dfrac{yyy}{yy} = y$.

And $a^{x+1} \div a = a^{n+1-1} = a^n$. That is $\dfrac{aa^n}{a} = a^n$.

And $x^n \div x^n = x^{n-n} = x^0 = 1$. That is $\dfrac{x^n}{x^n} = 1$.

Divide	y^{2m}	b^6	$8a^{n+m}$	a^{n+3}	$12(b+y)^n$
By	y^m	b^3	$4a^m$	a^2	$3(b+y)^3$
Quot.	y^2		$2a^n$		$4(b+y)^{n-3}$

238. The rule is equally applicable to powers whose exponents are *negative*.

1. The quotient of a^{-5} by a^{-3}, is a^{-2}.

That is $\dfrac{1}{aaaaa} \div \dfrac{1}{aaa} = \dfrac{1}{aaaaa} \times \dfrac{aaa}{1} = \dfrac{aaa}{aaaaa} = \dfrac{1}{aa}$.

2. $-x^{-5} \div x^{-3} = \div x^{-2}$. That is $\dfrac{1}{-x^5} \div \dfrac{1}{x^3} = \dfrac{x^3}{-x^5} = \dfrac{1}{-x^2}$,

3. $h^2 \div h^{-1} = h^{2+1} = h^3$. That is $h^2 \div \dfrac{1}{h} = h^2 \times \dfrac{h}{1} = h^3$.

In this example, -1 the index of the divisor is to be subtracted from $+2$, the index of the dividend. But -1 becomes by subtraction $+1$. (Art. 82.)

4. $6a^n \div 2a^{-3} = 3a^{n+3}$.   5. $ba^3 \div a = ba^2$.

6. $b^3 \div b^5 = b^{3-5} = b^{-2}$.   7. $a^4 \div a^7 = a^{-3}$.

8. $(a^3 + y^3)^m \div (a^3 + y^3)^n = (a^3 + y^3)^{m-n}$.

9. $(b+x)^n \div (b+x) = (b+x)^{n-1}$.

The multiplication and division of powers by adding and subtracting their indices, should be made very familiar; as they have numerous and important applications, in the higher branches of algebra.

EXAMPLES OF FRACTIONS CONTAINING POWERS.

239. In the section on fractions, the following examples were omitted, for the sake of avoiding an anticipation of the subject of powers.

1. Reduce $\dfrac{5a^4}{3a^2}$ to lower terms.   Ans, $\dfrac{5a^2}{3}$.

For $\dfrac{5a^4}{3a^2} = \dfrac{5aaaa}{3aa} = \dfrac{5aa}{3}$.   (Art. 145.)

2. Reduce $\dfrac{6x^6}{3x^5}$ to lower terms.   Ans. $\dfrac{2x}{1}$ or $2x$. ·

3. Reduce $\dfrac{3a^4+4a^6}{5a^3}$ to lower terms.   Ans. $\dfrac{3a+4a^3}{5}$.

4. Reduce $\dfrac{8a^3y-12a^2y^2+6ay^3}{6a^2y+4ay^2}$ to lower terms.

Ans. $\dfrac{4a^2-6ay+3y^2}{3a+2y}$ obtained by dividing each term by $2ay$.

5. Reduce $\dfrac{a^2}{a^3}$, and $\dfrac{a^{-3}}{a^{-4}}$, to a common denominator.

$a^2 \times a^{-4}$ is a^{-2}, the first numerator. (Art. 146.)
$a^3 \times a^{-3}$ is $a^0 = 1$, the second numerator.
$a^3 \times a^{-4}$ is a^{-1}, the common denominator.

The fractions reduced are therefore $\dfrac{a^{-2}}{a^{-1}}$ and $\dfrac{1}{a^{-1}}$.

6. Reduce $\dfrac{2a^4}{5a^3}$ and $\dfrac{a^2}{a^4}$, to a common denominator.

·Ans. $\dfrac{2a^8}{5a^7}$ and $\dfrac{5a^5}{5a^7}$, or $\dfrac{2a^3}{5a^2}$ and $\dfrac{5}{5a^2}$. (Art. 145.)

7. Multiply $\dfrac{3x^2}{4x^3}$ into $\dfrac{dx}{2x^4}$.   Ans. $\dfrac{3dx^3}{8x^7} = \dfrac{3d}{8x^4}$.

8. Multiply $\dfrac{a^3+b}{b^4}$, into $\dfrac{a-b^2}{\cdot 3}$. ·

9. Multiply $\dfrac{a^5+1}{x^2}$, into $\dfrac{b^2-1}{x+a}$.

10. Multiply $\dfrac{b^4}{a^{-2}}$, into $\dfrac{h^{-3}}{x}$, and $\dfrac{a^n}{y^{-2}}$.

11. Divide $\dfrac{a^4}{y^3}$ by $\dfrac{a^3}{y^2}$.   Ans. $\dfrac{a^4}{y^3} \times \dfrac{y^2}{a^3} = \dfrac{a^4y^2}{a^3y^3} = \dfrac{a}{y}$.

12. Divide $\dfrac{a^3-x^4}{a^2}$, by $\dfrac{x^2-a^{-2}}{a}$.

13. Divide $\dfrac{b-y^{-1}}{y}$, by $\dfrac{a^3+b^{-4}}{y^3}$.

14. Divide $\dfrac{h^3-1}{d^4}$, by $\dfrac{d^n+1}{h}$.

SECTION IX.

EVOLUTION AND RADICAL QUANTITIES.[*]

ART. 240. IF a quantity is multiplied into itself, the product is a *power.* On the contrary, if a quantity is resolved into any number of *equal factors,* each of these is a *root* of that quantity.

Thus b is the root of bbb; because bbb may be resolved into the three equal factors b, and b, and b.

In subtraction, a quantity is resolved into *two parts.*

In division, a quantity is resolved into *two factors.*

In evolution, a quantity is resolved into *equal factors.*

241. *A root of a quantity, then, is a factor which multiplied into itself a certain number of times will produce that quantity.*

The number of times the root must be taken as a factor, to produce the given quantity, is denoted by the name of the root.

Thus 2 is the 4th root of 16; because $2 \times 2 \times 2 \times 2 = 16$, where 2 is taken *four* times as a factor, to produce 16.

So a^3 is the square root of a^6; for $a^3 \times a^3 = a^6$. (Art. 233.)

And a^2 is the cube root of a^6; for $a^2 \times a^2 \times a^2 = a^6$.

And a is the 6th root of a^6; for $a \times a \times a \times a \times a \times a = a^6$.

Powers and roots are correlative terms. If one quantity is a power of another, the latter is a root of the former. As b^3 is the cube of b; b is the cube root of b^x. As 9 is the square of 3; 3 is the square root of 9.

242. There are two methods in use, for expressing the roots of quantities, one by means of the radical sign $\sqrt{}$, and the other by a fractional index. The latter is generally to be preferred. But the former has its uses on particular occasions.

When a root is expressed by the radical sign, the sign is placed over the given quantity, in this manner $\sqrt{a}$.

---

[*] Newton's Arithmetic, Maclaurin, Emerson, Euler, Saunderson, and Simpson.

Thus $^2\sqrt{a}$ is the 2d or square root of a.

$^3\sqrt{a}$ is the 3d or cube root.

$^n\sqrt{a}$ is the nth root.

And $^n\sqrt{a+y}$ is the nth root of $a+y$.

243. The figure placed over the radical sign, denotes the number of factors into which the given quantity is resolved; in other words, the number of times the root must be taken as a factor, to produce the given quantity.

So that $^2\sqrt{a} \times {}^2\sqrt{a} = a$.

And $^3\sqrt{a} \times {}^3\sqrt{a} \times {}^3\sqrt{a} = a$.

And $^n\sqrt{a} \times {}^n\sqrt{a} \dots n$ times $= a$.

The figure for the *square* root is commonly omitted; $\sqrt{a}$ being put for $^2\sqrt{a}$. Whenever, therefore, the radical sign is used without a figure, the square root is to be understood.

244. When a figure or letter is *prefixed* to the radical sign, without any character between them; the two quantities are to be considered as *multiplied* together.

Thus $2\sqrt{a}$, is $2 \times \sqrt{a}$, that is, 2 multiplied into the root of a, or which is the same thing, *twice* the root of a.

And $x\sqrt{b}$, is $x \times \sqrt{b}$, or x times the root of b.

When no co-efficient is prefixed to the radical sign, 1 is always to be understood; $\sqrt{a}$ being the same as $1\sqrt{a}$, that is, *once* the root of a.

245. The method of expressing roots by radical signs, has no very apparent connection with the other parts of the scheme of algebraic notation. But the plan of indicating them by *fractional indices*, is derived directly from the mode of expressing *powers* by *integral* indices. To explain this, let a^6 be a given quantity. If the index be divided into any number of equal parts, each of these will be the index of a root of a^6.

Thus the *square* root of a^6, is a^3. For, according to the definition, (Art. 241.) the square root of a^6 is a factor, which multiplied into itself will produce a^6. But $a^3 \times a^3 = a^6$. (Art. 233.) Therefore, a^3 is the square root of a^6. The index of the given quantity a^6, is here divided into the two equal parts 3 and 3. Of course, the quantity itself is resolved into the two equal factors a^3 and a^3.

The *cube* root of a^6 is a^2. For $a^2 \times a^2 \times a^2 = a^6$.

Here the index is divided into *three* equal parts, and the quantity itself resolved into three equal factors.

The square root of a^2 is a^1 or a. For $a \times a = a^2$.

By extending the same plan of notation, *fractional indices* are obtained.

Thus, in taking the square root of a^1 or a, the index 1 is divided into the two equal parts $\frac{1}{2}$ and $\frac{1}{2}$; and the root is $a^{\frac{1}{2}}$.

On the same princple,

The cube root of a, is $a^{\frac{1}{3}} = {}^2\sqrt{a}$

The fourth root is $a^{\frac{1}{4}} = {}^4\sqrt{a}$

The nth root, is $a^{\frac{1}{n}} = {}^n\sqrt{a}$, &c.

And the nth root of $a+x$, is $(a+x)^{\frac{1}{n}} = {}^n\sqrt{a+x}$.

246. In all these cases, the denominator of the fractional index, expresses the number of factors into which the given quantity is resolved.

So that $a^{\frac{1}{2}} \times a^{\frac{1}{2}} = a$.

$a^{\frac{1}{3}} \times a^{\frac{1}{3}} \times a^{\frac{1}{3}} = a$.

$a^{\frac{1}{n}} \times a^{\frac{1}{n}} \ldots n$ times $= a$. See art. 243.

247. It follows from this plan of notation, that

$a^{\frac{1}{2}} \times a^{\frac{1}{2}} = a^{\frac{1}{2}+\frac{1}{2}}$. For $a^{\frac{1}{2}+\frac{1}{2}} = a^1$ or a.

$a^{\frac{1}{3}} \times a^{\frac{1}{3}} \times a^{\frac{1}{3}} = a^{\frac{1}{3}+\frac{1}{3}+\frac{1}{3}} = a^1$, &c.

where the multiplication is performed in the same manner, as the multiplication of powers, (Art. 233,) that is, by *adding the indices.*

248. Every root as well as every power of 1 is 1. (Art. 209.) For a root is a factor which multiplied into itself will produce the given quantity. But no factor except 1 can produce 1, by being multiplied into itself.

So that 1^n, 1, $\sqrt{1}$, ${}^n\sqrt{1}$, &c. are all equal.

249. *Negative* indices are used in the notation of roots, as well as of powers. See art. 207.

Thus $\dfrac{1}{a^{\frac{1}{2}}} = a^{-\frac{1}{2}}$ $\qquad \dfrac{1}{a^{\frac{1}{4}}} = a^{-\frac{1}{4}}$

$\dfrac{1}{a^{\frac{1}{3}}} = a^{-\frac{1}{3}}$ $\qquad \dfrac{1}{a^{\frac{1}{n}}} = a^{-\frac{1}{n}}$.

P

Powers of Roots.

250. It has been shewn in what manner any power or root may be expressed by means of an index. The index of a power is a whole number. That of a root is a fraction whose numerator is 1. There is also another class of quantities, which may be considered, either as powers of roots, or roots of powers.

Suppose $a^{\frac{1}{2}}$ is multiplied into itself, so as to be repeated three times as a factor.

The product $a^{\frac{1}{2}+\frac{1}{2}+\frac{1}{2}}$ or $a^{\frac{3}{2}}$ (Art. 247.) is evidently the cube of $a^{\frac{1}{2}}$, that is, the cube of the square root of a. This fractional index denotes, therefore, *a power of a root*. The denominator expresses the root, and the numerator the power. The denominator shows into how many equal factors or roots the given quantity is resolved; and the numerator shows how many of these roots are to be multiplied together.

Thus $a^{\frac{4}{3}}$ is the 4th power of the cube root of a.

The denominator shows that a is resolved into the three factors or roots $a^{\frac{1}{3}}$, and $a^{\frac{1}{3}}$, and $a^{\frac{1}{3}}$. And the numerator shows that four of these are to be multiplied together; which will produce the fourth power of $a^{\frac{1}{3}}$; that is,

$$a^{\frac{1}{3}} \times a^{\frac{1}{3}} \times a^{\frac{1}{3}} \times a^{\frac{1}{3}} = a^{\frac{4}{3}}.$$

251. As $a^{\frac{3}{2}}$ is a power of a root, so it is *a root of a power*. Let a be raised to the third power a^3. The square root of this is $a^{\frac{3}{2}}$. For the root of a^3 is a quantity which multiplied into itself will produce a^3.

But according to art. 247, $a^{\frac{3}{2}} = a^{\frac{1}{2}} \times a^{\frac{1}{2}} \times a^{\frac{1}{2}}$; and this multiplied into itself (Art. 103.) is

$$a^{\frac{1}{2}} \times a^{\frac{1}{2}} \times a^{\frac{1}{2}} \times a^{\frac{1}{2}} \times a^{\frac{1}{2}} \times a^{\frac{1}{2}} = a^3.$$

Therefore $a^{\frac{3}{2}}$ is the square root of the cube of a.

In the same manner, it may be shown that $a^{\frac{m}{n}}$ is the mth power of the nth root of a; or the nth root of the mth power: that is, *a root of a power is equal to the same power of the same root.* For instance, the fourth power of the cube root of a, is the same, as the cube root of the fourth power of a.

252. Roots, as well as powers, of the same letter may be multiplied by *adding their exponents.* (Art. 247.) It will be easy to see, that the same principle may be extended to powers of roots, when the exponents have a common denominator.

Thus $a^{\frac{2}{7}} \times a^{\frac{3}{7}} = a^{\frac{2}{7} + \frac{3}{7}} = a^{\frac{5}{7}}$.

For the first numerator shows how often $a^{\frac{1}{7}}$ is taken as a factor to produce $a^{\frac{2}{7}}$. (Art. 250.)

And the second numerator shows how often $a^{\frac{1}{7}}$ is taken as a factor to produce $a^{\frac{3}{7}}$.

The *sum* of the numerators, therefore, shows how often the root must be taken, for the *product.* (Art. 103.)

Or thus, $a^{\frac{2}{7}} = a^{\frac{1}{7}} \times a^{\frac{1}{7}}$.

And $a^{\frac{3}{7}} = a^{\frac{1}{7}} \times a^{\frac{1}{7}} \times a^{\frac{1}{7}}$.

Therefore $a^{\frac{2}{7}} \times a^{\frac{3}{7}} = a^{\frac{1}{7}} \times a^{\frac{1}{7}} \times a^{\frac{1}{7}} \times a^{\frac{1}{7}} \times a^{\frac{1}{7}} = a^{\frac{5}{7}}$.

253. The value of a quantity is not altered, by applying to it a fractional index whose numerator and denominator are equal.

Thus $a = a^{\frac{2}{2}} = a^{\frac{3}{3}} = a^{\frac{n}{n}}$. For the denominator shows that a is resolved into a certain number of factors; and the numerator shows that all these factors are included in $a^{\frac{n}{n}}$.

Thus $a^{\frac{2}{2}} = a^{\frac{1}{2}} \times a^{\frac{1}{2}}$, which is equal to a. (Art. 246.)

And $a^{\frac{3}{3}} = a^{\frac{1}{3}} \times a^{\frac{1}{3}} \times a^{\frac{1}{3}}$, which is also equal to a.

And $a^{\frac{n}{n}} = a^{\frac{1}{n}} \times a^{\frac{1}{n}} \times a^{\frac{1}{n}} \dots n$ times.

On the other hand, when the numerator of a fractional index becomes equal to the denominator, the expression may be rendered more simple by *rejecting* the index.

Instead of $a^{\frac{n}{n}}$, we may write a.

254. The index of a power or root may be exchanged, for any other index of the same value.

Instead of $a^{\frac{2}{3}}$, we may put $a^{\frac{4}{6}}$.

For, in the latter of these expressions, a is supposed to be resolved into *twice* as many factors as in the former; and the numerator shews that *twice* as many of these factors are to be multiplied together. So that the whole value is not altered.

The one is $\quad a^{\frac{1}{3}} \times a^{\frac{1}{3}} = a^{\frac{2}{3}}$.

The other is $a^{\frac{1}{6}} \times a^{\frac{1}{6}} \times a^{\frac{1}{6}} \times a^{\frac{1}{6}} = a^{\frac{4}{6}}$.

On the same principle $a^{\frac{2}{3}} = a^{\frac{2n}{3n}}$.

Thus $x^{\frac{2}{3}} = x^{\frac{4}{6}} = x^{\frac{6}{9}}$, &c. that is, the square of the cube root is the same, as the fourth power of the sixth root, the sixth power of the 9th root, &c.

So $a^2 = a^{\frac{4}{2}} = a^{\frac{6}{3}} = a^{\frac{2n}{n}}$. For the value of each of these indices is 2. (Art. 135.)

255. From the preceding article, it will be easily seen, that a fractional index may be expressed in *decimals*.

1. Thus $a^{\frac{1}{2}} = a^{\frac{5}{10}}$ or $a^{0.5}$; that is, the square root is equal to the 5th power of the tenth root.

2. $a^{\frac{1}{4}} = a^{\frac{25}{100}}$ or $a^{0.25}$; that is, the fourth root is equal to the 25th power of the 100th root.

3. $a^{\frac{2}{5}} = a^{0.4}$.

4. $a^{\frac{7}{2}} = a^{3.5}$.

5. $a^{\frac{9}{6}} = a^{1.5}$.

6. $a^{\frac{11}{4}} = a^{2.75}$.

In many cases however the decimal can be only an *approximation* to the true index.

Thus $a^{\frac{1}{3}} = a^{0.3}$ nearly.

$a^{\frac{1}{3}} = a^{0.33}$ more nearly.

$a^{\frac{1}{3}} = a^{0.33333}$ very nearly.

In this manner, the approximation may be carried to any degree of exactness which is required.

Thus $a^{\frac{5}{3}} = a^{1.66666}$.

$a^{\frac{5}{6}} = a^{0.83333}$.

$a^{\frac{11}{7}} = a^{1.57142}$.

These decimal indices form a very important class of numbers, called *logarithms*.

It is frequently convenient to vary the notation of powers of roots, by making use of a vinculum, or the radical sign $\sqrt{}$. In doing this, we must keep in mind, that the power of a root is the same, as the root of a power; (Art. 251.) and also, that the *denominator* of a fractional exponent expresses a *root*, and the *numerator*, a *power*. (Art. 250.)

Instead, therefore, of $a^{\frac{2}{3}}$, we may write $(a^{\frac{1}{3}})^2$, or $(a^2)^{\frac{1}{3}}$, or $\sqrt[3]{a^2}$.

The first of these three forms, denotes the square of the cube root of a; and each of the two last, the cube root of the square of a.

So $a^{\frac{m}{n}} = \overline{a^{\frac{1}{n}}}\Big|^m = \overline{a^m}\Big|^{\frac{1}{n}} = \sqrt[n]{a^m}$

And $(bx)^{\frac{3}{4}} = (b^3 x^3)^{\frac{1}{4}} = \sqrt[4]{b^3 x^3}$.

And $\overline{a+y}\Big|^{\frac{3}{5}} = \overline{\overline{a+y}\big|^3}\Big|^{\frac{1}{5}} = \sqrt[5]{\overline{a+y}^3}$.

EVOLUTION.

257. Evolution is the opposite of involution. One is finding a *power* of a quantity, by multiplying it into itself. The other is finding a *root*, by resolving a quantity into equal factors. A quantity is resolved into any number of equal factors, by dividing its *index* into as many *equal parts*. (Art. 245.)

Evolution may be performed, then, by the following general rule;

Divide the index of the quantity, by the number expressing the root to be found.

Or, place over the quantity the radical sign belonging to the required root.

1. Thus the cube root of a^6 is a^2. For $a^2 \times a^2 \times a^2 = a^6$.

Here 6, the index of the given quantity, is divided by 3, the number expressing the cube root.

2. The cube root of a or a^1, is $a^{\frac{1}{3}}$ or $\sqrt[3]{a}$.

For $a^{\frac{1}{3}} \times a^{\frac{1}{3}} \times a^{\frac{1}{3}}$, or $\sqrt[3]{a} \times \sqrt[3]{a} \times \sqrt[3]{a} = a$. (Arts. 213, 246.)

Here the index 1 is divided by 3.

3. The 5th root of ab, is $(ab)^{\frac{1}{5}}$ or $\sqrt[5]{ab}$.

Q

4. The nth root of a^2, is $a^{\frac{2}{n}}$ or $\sqrt[n]{a^2}$.

5. The 7th root of $2d-x$, is $(2d-x)^{\frac{1}{7}}$ or $\sqrt[7]{2d-x}$.

6. The 5th root of $\overline{a-x}|^3$, is $\overline{a-x}|^{\frac{3}{5}}$ or $\sqrt[5]{\overline{a-x}|^3}$.

7. The cube root of $a^{\frac{1}{2}}$, is $a^{\frac{1}{6}}$. (Art. 163.)

8. The 4th root of a^{-1} is $a^{-\frac{1}{4}}$

9. The cube root of $a^{\frac{2}{3}}$, is $a^{\frac{2}{9}}$.

10. The nth root of x^m, is $x^{\frac{m}{n}}$.

258. According to the rule just given, the cube root of the square root is found, by dividing the index $\frac{1}{2}$ by 3, as in example 7th. But instead of dividing by 3, we may *multiply* by $\frac{1}{3}$. For $\frac{1}{2} \div 3 = \frac{1}{2} \div \frac{3}{1} = \frac{1}{2} \times \frac{1}{3}$. (Art. 162.)

So $\dfrac{1}{m} \div n = \dfrac{1}{m} \times \dfrac{1}{n}$. Therefore the mth root of the nth root of a is equal to $a^{\frac{1}{n} \times \frac{1}{m}}$.

That is $\overline{a^{\frac{1}{n}}}|^{\frac{1}{m}} = a^{\frac{1}{n} \times \frac{1}{m}} = a^{\frac{1}{nm}}$

Here the two fractional indices are reduced to one by multiplication.

It is sometimes necessary to *reverse* this process; to re-solve an index into *two factors*.

Thus $x^{\frac{1}{8}} = x^{\frac{1}{4} \times \frac{1}{2}} = \overline{x^{\frac{1}{4}}}|^{\frac{1}{2}}$. That is, the 8th root of x is equal to the square root of the 4th root.

So $\overline{a+b}|^{\frac{1}{mn}} = \overline{a+b}|^{\frac{1}{m} \times \frac{1}{n}} = \overline{\overline{a+b}|^{\frac{1}{m}}}|^{\frac{1}{n}}$.

It may be necessary to observe, that resolving the *index* into factors, is not the same as resolving the *quantity* into factors. The latter is effected, by dividing the index into *parts*.

259. The rule in art. 257, may be applied to every case in evolution. But when the quantity whose root is to be found, is composed of *several factors*, there will frequently be an advantage in taking the root of each of the factors separately.

This is done upon the principle, that *the root of the product of several factors, is equal to the product of their roots.*

Thus $\sqrt{ab}=\sqrt{a}\times\sqrt{b}$. For each member of the equation, if involved, will give the same power.

The square of $\sqrt{ab}$ is ab. (Art. 241.)

The square of $\sqrt{a}\times\sqrt{b}$, is $\sqrt{a}\times\sqrt{a}\times\sqrt{b}\times\sqrt{b}$. (Art 102.)

But $\sqrt{a}\times\sqrt{a}=a$. (Art. 241.) And $\sqrt{b}\times\sqrt{b}=b$.

Therefore the square of $\sqrt{a}\times\sqrt{b}=\sqrt{a}\times\sqrt{a}\times\sqrt{b}\times\sqrt{b}=ab$, which is also the square of $\sqrt{ab}$.

On the same principle, $(ab)^{\frac{1}{n}} = a^{\frac{1}{n}} b^{\frac{1}{n}}$.

When, therefore, a quantity consists of several factors, we may either extract the root of the whole together; or we may find the root of the factors separately, and then multiply them into each other.

Ex. 1. The cube root of xy, is either $(xy)^{\frac{1}{3}}$, or $x^{\frac{1}{3}} y^{\frac{1}{3}}$.

2. The 5th root of $3y$, is $^5\sqrt{3y}$ or $^5\sqrt{3}\times{}^5\sqrt{y}$.

3. The 6th root of abh, is $(abh)^{\frac{1}{6}}$, or $a^{\frac{1}{6}} b^{\frac{1}{6}} h^{\frac{1}{6}}$.

4. The cube root of $8b$, is $(8b)^{\frac{1}{3}}$, or $2b^{\frac{1}{3}}$.

5. The nth root of $x^n y$, is, $(x^n y)^{\frac{1}{n}}$ or $xy^{\frac{1}{n}}$.

260. *The root of a fraction is equal to the root of the numerator divided by the root of the denominator.*

1. Thus the square root of $\dfrac{a}{b}=\dfrac{a^{\frac{1}{2}}}{b^{\frac{1}{2}}}$. For $\dfrac{a^{\frac{1}{2}}}{b^{\frac{1}{2}}}\times\dfrac{a^{\frac{1}{2}}}{b^{\frac{1}{2}}}=\dfrac{a}{b}$.

2. So the nth root of $\dfrac{a}{b}=\dfrac{a^{\frac{1}{n}}}{b^{\frac{1}{n}}}$. For $\dfrac{a^{\frac{1}{n}}}{b^{\frac{1}{n}}}\times\dfrac{a^{\frac{1}{n}}}{b^{\frac{1}{n}}}\cdots n$ times$=\dfrac{a}{b}$.

3. The square root of $\dfrac{x}{ay}$, is $\dfrac{^t\sqrt{x}}{\sqrt{ay}}$.

4. $\sqrt{\dfrac{ah}{xy}}=\dfrac{\sqrt{ah}}{\sqrt{xy}}$.

261. For determining what *sign* to prefix to a root, it is important to observe, that

An odd root of any quantity has the same sign as the quantity itself;

An even root of an affirmative quantity is ambiguous;

An even root of a negative quantity is impossible.

That the 3d, 5th, 7th, or any other *odd* root of a quantity, must have the same sign as the quantity itself, is evident from art. 219.

262. But an *even* root of an *affirmative* quantity, may be either affirmative or negative. For the quantity may be produced from the one, as well as from the other. (Art. 219.)

Thus the square root of a^2 is $+a$ or $-a$.

An even root of an affirmative quantity is, therefore said to be *ambiguous*, and is marked with both $+$ and $-$.

Thus the square root of $3b$, is $\pm\sqrt{3b}$.

The 4th root of x, is $\pm x^{\frac{1}{4}}$.

263. But no even root of a *negative* quantity can be found. Thus the square root of $-a^2$ is neither $+a$ nor $-a$.

For $+a \times +a = +a^2$.   And $-a \times -a = +a^2$ also.

An even root of a negative quantity is, therefore, said to be *impossible* or *imaginary*.

There are purposes to be answered, however, by applying the radical sign to negative quantities.* The expresion $\sqrt{-a}$ is often to be found in algebraic processes. For, although we are unable to assign it a rank, among either positive or negative quantities; yet we know that when multiplied into itself its product is $-a$, because $\sqrt{-a}$ is by notation a *root* of $-a$, that is, a quantity which multiplied into itself produces $-a$.

This may, at first view, seem to be an exception to the general rule that the product of two negatives is affirmative. But it is to be considered, that $\sqrt{-a}$ is not itself a negative quantity, but the *root* of a negative quantity.

It ought also to be observed that $\sqrt{-a}$ is not equivalent to $-\sqrt{a}$. The first is a root of $-a$, but the latter is a root of $+a$.

For $-\sqrt{a} \times -\sqrt{a} = +a$.

* See an interesting Essay, on the use of impossible quantities in calculation, by Professor Playfair, in the London Philosophical Transactions, for 1778.

264. The methods of extracting the roots of *compound* quantities are to be considered in a future section. But there is one class of these, the squares of *binomial* and *residual* quantities, which it will be proper to attend to in this place. It has been shown, (Art. 214.) that the square of a binomial quantity consists of *three terms*, two of which are complete powers, and the other is a double product of the roots of these powers. The square of $a+b$, for instance, is

$$a^2 + 2ab + b^2,$$

two terms of which, a^2 and b^2, are complete powers, and $2ab$ is twice the product of a into b, that is, of the root of a^2 into the root of b^2.

Whenever, therefore, we meet with a quantity of this description, we may know that its square root is a binomial ; and this may be found, by taking the root of the two terms which are complete powers, and connecting them by the sign $+$. The other term disappears in the root. Thus to find the square root of

$$x^2 + 2xy + y^2,$$

take the root of x^2, and the root of y^2, and connect them by the sign $+$. The binomial root will then be $x+y$.

In a *residual* quantity, the double product has the sign $-$ prefixed, instead of $+$. The square of $a-b$, for instance, is $a^2 - 2ab + b^2$. (Art. 214.) And to obtain the root of a quantity of this description, we have only to take the roots of the two complete powers, and connect them by the sign $-$. Thus the square root of $x^2 - 2xy + y^2$ is $x-y$. Hence,

265. To extract a binomial or residual square root, *take the roots of the two terms which are complete powers, and connect them by the sign, which is prefixed to the other term.*

Ex. 1. Find the root of $x^2 + 2x + 1$.
The two terms which are complete powers are x^2 and 1. Their roots are x and 1. (Art. 248.)
The binomial root is, therefore, $x+1$.

2. The square root of $x^2 - 2x + 1$, is $x-1$. (Art. 214.)

3. The square root of $a^2 + a + \frac{1}{4}$, is $a + \frac{1}{2}$. (Art. 224.)

4. The square root of $a^2 + \frac{4}{3} a + \frac{4}{9}$, is $a + \frac{2}{3}$.

5. The square root of $a^2 + ab + \frac{b^2}{4}$, is $a + \frac{b}{2}$.

6. The square root of $a^2 + \frac{2ab}{c} + \frac{b^2}{c^2}$, is $a + \frac{b}{c}$.

R

266. *A root whose value cannot be exactly expressed in numbers, is called a* SURD.

Thus $\sqrt{2}$ is a surd, because the square root of 2 cannot be expressed in numbers, with perfect exactness.

In decimals, it is 1.41421356 nearly.

But though we are unable to assign the value of such a quantity *when taken alone*, yet by multiplying it into itself, or by combining it with other quantities, we may produce expressions whose value can be determined. There is therefore a system of rules generally appropriated to surds. But as all quantities whatever, when under the same radical sign, or having the same index, may be treated in nearly the same manner; it will be most convenient to consider them together, under the general name of *Radical Quantities ;* understanding by this term, every quantity which is found under a radical sign, or which has a fractional index.

267. Every quantity which is not a surd, is said to be *rational.* But for the purpose of distinguishing between radicals and other quantities, the term rational will be applied, in this section, to those only which do not appear under a radical sign, and which have not a fractional index.

REDUCTION OF RADICAL QUANTITIES.

268. Before entering on the consideration of the rules for the addition, subtraction, multiplication, and division of radical quantities, it will be necessary to attend to the methods of reducing them from one form to another.

First, to reduce a *rational* quantity to the form of a radical;

Raise the quantity to a power of the same name as the given root, and then apply the corresponding radical sign or index.

Ex. 1. Reduce a to the form of the nth root.

The nth power of a is a^n. (Art. 211.)

Over this place the radical sign, and it becomes $\sqrt[n]{a^n}$.

It is thus reduced to the form of a radical quantity, without any alteration of its value. For $\sqrt[n]{a^n} = a^{\frac{n}{n}} = a.$ (Art. 253.)

2. Reduce 4 to the form of the cube root.

Ans. $\sqrt[3]{64}$ or $(64)^{\frac{1}{3}}$

c

3. Reduce $3a$ to the form of the 4th root.

Ans. $\sqrt[4]{81a^4}$.

4. Reduce $\frac{1}{3}ab$ to the form of the square root.

Ans. $(\frac{1}{9}a^2b^2)^{\frac{1}{2}}$.

5. Reduce $3 \times \overline{a-x}$ to the form of the cube root.

Ans. $\sqrt[3]{27 \times \overline{a-x}|^3}$. See art. 212.

6. Reduce a^2 to the form of the cube root.
The cube of a^2 is a^6. (Art. 220.)

And the cube root of a^6 is $\sqrt[3]{a^6} = a^{6\overline{|\frac{1}{3}}}$.

In cases of this kind, where a *power* is to be reduced to the form of the nth root, it must be raised to the nth power, not of the *given letter*, but of *the power* of the letter.

Thus in the example, a^6 is the cube, not of a, but of a^2.

7. Reduce a^3b^4 to the form of the square root.

Ans. $\sqrt{a^6b^8}$, or $(a^6b^8)^{\frac{1}{2}}$.

8. Reduce a^m to the form of the nth root.

Ans. $a^{mn\overline{|^{\frac{1}{a}}}}$.

269. *Secondly,* to reduce quantities which have different indices, to others of the same value having a *common index;*
1. Reduce the indices to a common denominator;
2. Involve each quantity, to the power expressed by the numerator of its reduced index.
3. Take the root denoted by the common denominator.

Ex. 1. Reduce $a^{\frac{1}{4}}$ and $b^{\frac{1}{6}}$ to a common index.
1st. The indices $\frac{1}{4}$ and $\frac{1}{6}$ reduced to a common denominator, are $\frac{3}{12}$ and $\frac{2}{12}$. (Art. 146.)
2d. The quantities a and b involved to the powers expressed by the two numerators, are a^3 and b^2.
3d. The root denoted by the common denominator is $\frac{1}{12}$.
The answer, then, is $a^3\overline{|^{\frac{1}{12}}}$ and $b^2\overline{|^{\frac{1}{12}}}$.

The two quantities are thus reduced to a common index, without any alteration in their values.

For by art. 254, $a^{\frac{1}{4}} = a^{\frac{3}{12}}$, which by art. 253, $= a^3\overline{|^{\frac{1}{12}}}$,
And universally $a^{\frac{1}{n}} = a^{\frac{m}{nm}} = a^m\overline{|^{\frac{1}{mn}}}$.

2. Reduce $a^{\frac{1}{2}}$ and $bx^{\frac{2}{3}}$ to a common index.

The indices reduced to a common denominator are $\frac{3}{6}$ and $\frac{4}{6}$.

The quantities, then, are $a^{\frac{3}{6}}$ and $(bx)^{\frac{4}{6}}$, or $\overline{a^3}|^{\frac{1}{6}}$ and $\overline{b^4 x^4}|^{\frac{1}{6}}$.

3. Reduce a^2 and $b^{\frac{1}{n}}$. Ans. $\overline{a^{2n}}|^{\frac{1}{n}}$ and $b^{\frac{1}{n}}$.

4. Reduce $x^{\frac{1}{n}}$ and $y^{\frac{1}{m}}$. Ans. $\overline{x^m}|^{\frac{1}{mn}}$ and $\overline{y^n}|^{\frac{1}{mn}}$.

5. Reduce $2^{\frac{1}{2}}$ and $3^{\frac{1}{3}}$. Ans. $8^{\frac{1}{6}}$ and $9^{\frac{1}{6}}$.

6. Reduce $(a+b)^2$ and $(x-y)^{\frac{2}{3}}$. Ans. $\overline{a+b}|^{6}|^{\frac{1}{3}}$ and $\overline{x-y}|^{2}|^{\frac{1}{3}}$.

270. When it is required to reduce a quantity to a *given* index;

Divide the index of the quantity by the given index, place the quotient over the quantity, and set the given index over the whole.

This is merely resolving the original index into two factors, according to art. 258.

Ex. 1. Reduce $a^{\frac{1}{6}}$ to the index $\frac{1}{2}$.

By art. 162, $\frac{1}{6} \div \frac{1}{2} = \frac{1}{6} \times \frac{2}{1} = \frac{2}{6} = \frac{1}{3}$.

This is the index to be placed over a, which then becomes $a^{\frac{1}{3}}$; and the given index set over this makes it $\overline{a^{\frac{1}{3}}}|^{\frac{1}{2}}$, the answer.

2. Reduce a^2 and $x^{\frac{3}{2}}$, to the common index $\frac{1}{3}$.

$2 \div \frac{1}{3} = 2 \times 3 = 6$, the first index $\bigg\}$
$\frac{3}{2} \div \frac{1}{3} = \frac{3}{2} \times 3 = \frac{9}{2}$, the second index

Therefore $(a^6)^{\frac{1}{3}}$ and $(x^{\frac{9}{2}})^{\frac{1}{3}}$ are the quantities required.

3. Reduce $4^{\frac{1}{2}}$ and $3^{\frac{1}{3}}$, to the common index $\frac{1}{6}$.

Answer. $(4^3)^{\frac{1}{6}}$ and $(3^2)^{\frac{1}{6}}$.

271. *Thirdly*, to remove a part of a root from under the radical sign;

If the quantity can be resolved into two factors, one of which is an exact power of the same name with the root; *find the root of this power, and prefix it to the other factor, with the radical sign between them.*

This rule is founded on the principle, that the root of the *product* of two factors is equal to the product of their roots. (Art. 259.)

It will generally be best to resolve the radical quantity into such factors, that one of them shall be the *greatest* power which will divide the quantity without a remainder. If there is no exact power which will divide the quantity, the reduction can not be made.

Ex. 1. Remove a factor from $\sqrt{8}$.

The greatest square which will divide 8 is 4.

We may then resolve 8 into the factors 4 and 2. For $4 \times 2 = 8$.

The root of this product is equal to the product of the roots of its factors; that is, $\sqrt{8} = \sqrt{4} \times \sqrt{2}$.

But $\sqrt{4} = 2$. Instead of $\sqrt{4}$, therefore, we may substitute its equal 2. We then have $2 \times \sqrt{2}$ or $2\sqrt{2}$.

This is commonly called reducing a radical quantity to its *most simple terms*. But the learner may not perhaps at once perceive, that $2\sqrt{2}$ is a more simple expression than $\sqrt{8}$.

2. Reduce $\sqrt{a^2 x}$. Ans. $\sqrt{a^2} \times \sqrt{x} = a \times \sqrt{x} = a\sqrt{x}$.

3. Reduce $\sqrt{18}$. Ans. $\sqrt{9 \times 2} = \sqrt{9} \times \sqrt{2} = 3\sqrt{2}$.

4. Reduce $\sqrt[3]{64 b^3 c}$. Ans. $\sqrt[3]{64 b^3} \times \sqrt[3]{c} = 4b^3\sqrt{c}$.

5. Reduce $\sqrt[4]{\dfrac{a^4 b}{c^5 d}}$. Ans. $\dfrac{a}{c} \sqrt[4]{\dfrac{b}{cd}}$. (Art. 260.)

6. Reduce $\sqrt[n]{a^n b}$. Ans. $a\sqrt[n]{b}$, or $ab^{\frac{1}{n}}$.

7. Reduce $(a^3 - a^2 b)^{\frac{1}{2}}$. Ans. $a(a-b)^{\frac{1}{2}}$.

8. Reduce $(54 a^6 b)^{\frac{1}{3}}$. Ans. $3a^2 (2b)^{\frac{1}{3}}$.

272. By a contrary process the co-efficient of a radical quantity may be introduced under the radical sign.

1. Thus $a\sqrt[n]{b} = \sqrt[n]{a^n b}$.

For $a = \sqrt[n]{a^n}$ or $a^{\frac{n}{n}}$. (Art. 253.) And $\sqrt[n]{a^n} \times \sqrt[n]{b} = \sqrt[n]{a^n b}$.

Here the co-efficient a is first raised to a power of the same name as the radical part, and is then introduced as a factor under the radical sign.

2. $a(x-b)^{\frac{1}{3}} = (a^3 \times \overline{x-b})^{\frac{1}{3}} = (a^3 x - a^3 b)^{\frac{1}{3}}$.

3. $2ab(2ab^2)^{\frac{1}{3}} = (16 a^4 b^5.)$

4. $\dfrac{a}{b}\left(\dfrac{b^2 c}{a^2+b^2}\right)^{\frac{1}{2}} = \left(\dfrac{a^2 b^2 c}{a^2 b^2+b^4}\right)^{\frac{1}{2}}.$

ADDITION AND SUBTRACTION OF RADICAL QUANTITIES.

273. Radical quantities may be added like rational quantities, *by writing them one after another with their signs.* (Art. 69.)

Thus the sum of $\sqrt{a}$ and $\sqrt{b}$, is $\sqrt{a}+\sqrt{b}.$

And the sum of $a^{\frac{1}{2}}-h^{\frac{1}{3}}$ and $x^{\frac{1}{4}}-y^{\frac{1}{n}}$, is $a^{\frac{1}{2}}-h^{\frac{1}{3}}+x^{\frac{1}{4}}-y^{\frac{1}{n}}$

But in many cases, several terms may be reduced to one, as in arts. 72 and 74.

The sum of $2\sqrt{a}$ and $3\sqrt{a}$ is $2\sqrt{a}+3\sqrt{a}=5\sqrt{a}.$

For it is evident that twice the root of a, and three times the root of a, are *five* times the root of a. Hence,

274. When the quantities to be added have the same radical part, under the same radical sign or index ; *add the rational parts, and to the sum annex the* RADICAL PARTS.

If no rational quantity is prefixed to the radical sign, 1 is always to be understood. (Art. 244.)

To	$2^h\sqrt{ay}$	$5\sqrt{a}$	$3(x+h)^{\frac{1}{7}}$	$5bh^{\frac{1}{6}}$	$a\sqrt{b-h}$
Add	$4\sqrt{ay}$	$-2\sqrt{a}$	$4(x+h)^{\frac{1}{7}}$	$7bh^{\frac{1}{6}}$	$y\sqrt{b-h}$
Sum	$3^h\sqrt{ay}$		$7(x+h)^{\frac{1}{7}}$		$(a+y)\times\sqrt{b-h}$

275. If the radical parts are originally different, they may sometimes be made alike, by the reductions in the preceding articles.

1. Add $\sqrt{8}$ to $\sqrt{50}$. Here the radical parts are not the same. But by the reduction in art. 271, $\sqrt{8}=2\sqrt{2}$, and $\sqrt{50}=5\sqrt{2}$. The sum then is $7\sqrt{2}.$

2. Add $\sqrt{16b}$ to $\sqrt{4b}$. Ans. $4\sqrt{b}+2\sqrt{b}=6\sqrt{b}.$

3. Add $\sqrt{a^2 x}$ to $\sqrt{b^4 x}$. Ans. $a\sqrt{x}+b^2\sqrt{x}=(a+b^2)\times\sqrt{x}.$

4. Add $(36a^2 y)^{\frac{1}{2}}$ to $(25y)^{\frac{1}{2}}$. Ans. $(6a+5)\times y^{\frac{1}{2}}.$

5. Add $\sqrt{18a}$ to $3\sqrt{2a}.$

276. But if the radical parts, after reduction, are *different,* or have different *exponents,* they cannot be united in the same term; and must be added by writing them one after the other.

The sum of $3\sqrt{b}$ and $2\sqrt{a}$, is $3\sqrt{b}+2\sqrt{a}$.

It is manifest that three times the root of b, and twice the root of a, are neither five times the root of b, nor five times the root of a, unless b and a are equal.

The sum of $^2\sqrt{a}$ and $^3\sqrt{a}$, is $^2\sqrt{a}+^3\sqrt{a}$.

The *square* root of a, and the *cube* root of a, are neither twice the square root, nor twice the cube root of a.

277. *Subtraction* of radical quantities is to be performed in the same manner as addition, except that the signs in the subtrahend are to be changed according to art. 82.

From	$\sqrt{ay}$	$4\,^n\sqrt{a+x}$	$3h^{\frac{1}{3}}$	$a(x+y)$	$-a^{-\frac{1}{n}}$
Sub.	$3\sqrt{ay}$	$3\,^n\sqrt{a+x}$	$-5h^{\frac{1}{3}}$	$b(x+y)$	$-2a^{-\frac{1}{n}}$
Diff.	$-2\sqrt{ay}$		$8h^{\frac{1}{3}}$		$a^{-\frac{1}{n}}$

From $\sqrt{50}$, subtract $\sqrt{8}$.. Ans. $5\sqrt{2}-2\sqrt{2}=3\sqrt{2}$.(Art.275.)
From $^3\sqrt{b^4y}$, subtract $^3\sqrt{by^4}$. Ans. $(b-y)\times{}^3\sqrt{by}$.
From $^n\sqrt{x}$, subtract $^5\sqrt{x}$.

MULTIPLICATION OF RADICAL QUANTITIES.

Radical quantities may be multiplied, like other quantities, by writing the factors one after another, either with or without the sign of multiplication between them. (Art. 93.)

Thus the product of $\sqrt{a}$ into $\sqrt{b}$, is $\sqrt{a}\times\sqrt{b}$.

The product of $h^{\frac{1}{3}}$ into $y^{\frac{1}{2}}$ is $h^{\frac{1}{3}}y^{\frac{1}{2}}$.

But it is often expedient to bring the factors under the same radical sign. This may be done, if they are first reduced to a common index.

Thus $^n\sqrt{x}\times{}^n\sqrt{y}={}^n\sqrt{xy}$. For the root of the product of several factors is equal to the product of their roots. (Art. 259.) Hence,

279. *Quantities under the same radical sign or index, may*

*be multiplied together like rational quantities, the product being placed under the common radical sign or index.**

Multiply $^2\sqrt{x}$ into $^3\sqrt{y}$, that is, $x^{\frac{1}{2}}$ into $y^{\frac{1}{3}}$.

The quantities reduced to the same index, (Art. 269.) are $(x^3)^{\frac{1}{6}}$, and $(y^2)^{\frac{1}{6}}$, and their product is $(x^3 y^2)^{\frac{1}{6}} = {}^6\sqrt{x^3 y^2}$.

Mult.	$\sqrt{a+m}$	$\sqrt{dx}$	$a^{\frac{3}{n}}$	$(a+y)^{\frac{1}{n}}$	$a^{\frac{1}{m}}$
Into,	$\sqrt{a-m}$	$\sqrt{hy}$	$x^{\frac{1}{2}}$	$(b+h)^{\frac{1}{n}}$	$x^{\frac{1}{n}}$
Prod.	$\sqrt{a^2-m^2}$		$(a^3 x)^{\frac{1}{2}}$		$(a^n x^m)^{\frac{1}{mn}}$

Multiply $\sqrt{8xb}$ into $\sqrt{2xb}$. Prod. $\sqrt{16x^2 b^2} = 4xb$.

In this manner the product of radical quantities often becomes *rational*.

Thus the product of $\sqrt{2}$ into $\sqrt{18} = \sqrt{36} = 6$.

And the product of $(a^2 y^3)^{\frac{1}{4}}$ into $(a^2 y)^{\frac{1}{4}} = (a^4 y^4)^{\frac{1}{4}} = ay$.

280. *Roots of the same letter or quantity may be multiplied, by adding their fractional exponents.*

The exponents, like all other fractions, must be reduced to a common denominator, before they can be united in one term. (Art. 148.)

Thus $a^{\frac{1}{2}} \times a^{\frac{1}{3}} = a^{\frac{1}{2}+\frac{1}{3}} = a^{\frac{3}{6}+\frac{2}{6}} = a^{\frac{5}{6}}$.

The values of the roots are not altered, by reducing their indices to a common denominator. (Art. 254.)

Therefore the first factor $a^{\frac{1}{2}} = a^{\frac{3}{6}}$ $\Big\}$
And the second $a^{\frac{1}{3}} = a^{\frac{2}{6}}$

But $a^{\frac{3}{6}} = a^{\frac{1}{6}} \times a^{\frac{1}{6}} \times a^{\frac{1}{6}}$. (Art. 250.)

And $a^{\frac{2}{6}} = a^{\frac{1}{6}} \times a^{\frac{1}{6}}$. [103, 250]

The product therefore is $a^{\frac{1}{6}} \times a^{\frac{1}{6}} \times a^{\frac{1}{6}} \times a^{\frac{1}{6}} \times a^{\frac{1}{6}} = a^{\frac{5}{6}}$.(Art.

And in all instances of this nature, the common denominator of the indices denotes a certain root, and the sum of

---

*The case of an *imaginary* root of a negative quantity is an exception. (Art. 263.)

the numerators shows how often this is to be repeated as a factor to produce the required product.

Thus $a^{\frac{1}{n}} \times a^{\frac{1}{m}} = a^{\frac{m}{mn}} \times a^{\frac{n}{mn}} = a^{\frac{m+n}{mn}}.$

Mult.	$3y^{\frac{1}{4}}$	$a^{\frac{1}{2}} \times a^{\frac{1}{3}}$	$(a+b)^{\frac{1}{2}}$	$(a-y)^{\frac{1}{n}}$	$x^{-\frac{1}{4}}$
Into	$y^{\frac{2}{3}}$	$a^{\frac{1}{4}}$	$(a+b)^{\frac{1}{4}}$	$(a-y)^{\frac{1}{m}}$	$x^{-\frac{1}{3}}$
Prod.	$3y^{1\frac{1}{2}}$		$(a+b)^{\frac{3}{4}}$		$x^{-1\frac{7}{12}}$

The product of $y^{\frac{1}{2}}$ into $y^{-\frac{1}{3}}$ is $y^{\frac{3}{6}-\frac{2}{6}} = y^{\frac{1}{6}}.$

Here the sum of the indices $\frac{3}{6}$ and $-\frac{2}{6}$ is $\frac{1}{6}$, according to the rule for reduction in addition. (Art. 74, or 148.).

Or thus, $y^{\frac{3}{6}} \times y^{-\frac{2}{6}} = y^{\frac{3}{6}} \times \dfrac{1}{y^{\frac{2}{6}}} = \dfrac{y^{\frac{3}{6}}}{y^{\frac{2}{6}}} = \dfrac{y^{\frac{1}{6}} \times y^{\frac{1}{6}} \times y^{\frac{1}{6}}}{y^{\frac{1}{6}} \times y^{\frac{1}{6}}} = y^{\frac{1}{6}}.$

The product of $a^{\frac{1}{n}}$ into $a^{-\frac{1}{n}}$, is $a^{\frac{1}{n}-\frac{1}{n}} = a^0 = 1.$

And $x^{n-\frac{1}{2}} \times x^{\frac{1}{2}-n} = x^{n-n+\frac{1}{2}-\frac{1}{2}} = x^0 = 1.$

The product of a^2 into $a^{\frac{1}{3}} = a^{\frac{6}{3}} \times a^{\frac{1}{3}} = a^{\frac{7}{3}}.$

281. From the last example, it will be seen, that *powers* and *roots* may be multiplied by a common rule. This is one of the many advantages derived from the notation by fractional indices. Any quantities whatever may be reduced to the form of radicals, (Art. 268.) and may then be subjected to the same modes of operation.

Thus $y^3 \times y^{\frac{1}{6}} = y^{3+\frac{1}{6}} = y^{\frac{19}{6}}$. [Art. 150.]

And $x \times x^{\frac{1}{n}} = x^{1+\frac{1}{n}} = x^{\frac{n+1}{n}}.$

The product will become rational, whenever the numerator of the index can be exactly divided by the denominator.

Thus $a^3 \times a^{\frac{1}{3}} \times a^{\frac{2}{3}} = a^{\frac{12}{3}} = a^4.$ (Art. 254.)

And $(a+b)^{\frac{4}{3}} \times (a+b)^{-\frac{1}{3}} = (a+b)^{\frac{3}{3}} = a+b.$

And $a^{\frac{3}{5}} \times a^{\frac{2}{5}} = a^{\frac{5}{5}} = a.$

282. When radical quantities which are reduced to the same index have *rational co-efficients, the rational parts may be multiplied together, and their product prefixed to the product of the radical parts.*

1. Multiply $a\sqrt{b}$ into $c\sqrt{d}$.

The product of the rational parts is ac.

The product of the radical parts is $\sqrt{bd}$. (Art. 279.)

And the whole product is $ac\sqrt{bd}$.

For $a\sqrt{b}$ is $a \times \sqrt{b}$. (Art. 244.) And $c\sqrt{d}$ is $c \times \sqrt{d}$.

By art. 102, $a \times \sqrt{b}$ into $c \times \sqrt{d}$, is $a \times \sqrt{b} \times c \times \sqrt{d}$; or by changing the order of the factors,

$$a \times c \times \sqrt{b} \times \sqrt{d} = ac \times \sqrt{bd} = ac\sqrt{bd}.$$

2. Multiply $ax^{\frac{1}{2}}$ into $bd^{\frac{1}{3}}$.

When the radical parts are reduced to a common index, the factors become $a(x^3)^{\frac{1}{6}}$ and $b(d^2)^{\frac{1}{6}}$.

The product then is $ab(x^3 d^2)^{\frac{1}{6}}$.

But in cases of this nature, we may save the trouble of reducing to a common index, by multiplying as in art. 278.

Thus $ax^{\frac{1}{2}}$ into $bd^{\frac{1}{3}}$, is $ax^{\frac{1}{2}}bd^{\frac{1}{3}}$.

Mult.	$a(b+x)^{\frac{1}{2}}$	$a\sqrt{y^2}$	$a\sqrt{x}$	$ax^{-\frac{1}{2}}$	$x \sqrt[3]{y^3}$
Into	$y(b-x)^{\frac{1}{2}}$	$b\sqrt{hy}$	$b\sqrt{x}$	$by^{-\frac{1}{2}}$	$y \sqrt[3]{9}$
Prod.	$ay(b^2-x^2)^{\frac{1}{2}}$		$ab\sqrt{x^2}=abx$		$3xy$

283. If the rational quantities, instead of being *co-efficients* to the radical quantities, are connected with them by the signs $+$ and $-$, each term in the multiplier must be multiplied into each in the multiplicand, as in art. 100.

Multiply $a+\sqrt{b}$
Into $\quad\quad c+\sqrt{d}$

$ac+c\sqrt{b}$ (Art. 244.)
$\quad a\sqrt{d}+\sqrt{bd}$ (Art. 279.)

$ac+c\sqrt{b}+a\sqrt{d}+\sqrt{bd}$

The product of $a+\sqrt{y}$ into $1+r\sqrt{y}$ is
$$a+\sqrt{y}+ar\sqrt{y}+r\sqrt{y^2} \text{ or } ry.$$

DIVISION OF RADICAL QUANTITIES.

284. The division of radical quantities may be expressed, by writing the divisor under the dividend, in the form of a fraction.

Thus the quotient of $\sqrt[3]{a}$ divided by $\sqrt{b}$, is $\dfrac{\sqrt[3]{a}}{\sqrt{b}}$.

And $(a+h)^{\frac{1}{3}}$ divided by $(b+x)^{\frac{1}{n}}$ is $\dfrac{(a+h)^{\frac{1}{3}}}{(b+x)^{\frac{1}{n}}}$.

In these instances, the radical sign or index is *separately* applied to the numerator and the denominator. But if the divisor and dividend are reduced to the *same* index or radical sign, this may be applied to the *whole* quotient.

Thus $\sqrt[n]{a} \div \sqrt[n]{b} = \dfrac{\sqrt[n]{a}}{\sqrt[n]{b}} = \sqrt[n]{\dfrac{a}{b}}$. For the root of a fraction is equal to the root of the numerator divided by the root of the denominator. [Art. 260.]

Again $\sqrt[n]{ab} \div \sqrt[n]{b} = \sqrt[n]{a}$. For the product of this quotient into the divisor is equal to the dividend, that is, $\sqrt[n]{a} \times \sqrt[n]{b} = \sqrt[n]{ab}$. [Art. 279.] Hence,

285. *Quantities under the same radical sign or index, may be divided like rational quantities, the quotient being placed under the common radical sign or index.*

Divide $(x^3 y^2)^{\frac{1}{6}}$ by $y^{\frac{1}{3}}$.

These reduced to the same index are $(x^3 y^2)^{\frac{1}{6}}$ and $(y^2)^{\frac{1}{6}}$. [Art. 269.]

And the quotient is $(x^3)^{\frac{1}{6}} = x^{\frac{3}{6}} = x^{\frac{1}{2}}$. [Art. 258.]

Divide	$\sqrt{6a^3x}$	$\sqrt{dhx^2}$	$(a^3+ax)^{\frac{1}{9}}$	$(a^3h)^{\frac{1}{m}}$	$(a^2y^2)^{\frac{1}{4}}$
By	$\sqrt{3x}$	$\sqrt{dx}$	$a^{\frac{1}{9}}$	$(ax)^{\frac{1}{m}}$	$(ay)^{\frac{1}{4}}$
Quot.	$\sqrt{2a^3}$		$(a^2+x)^{\frac{1}{9}}$		$(ay)^{\frac{1}{4}}$

286. *A root is divided by another root of the same letter or quantity, by subtracting the index of the divisor from that of the dividend.*

Thus $a^{\frac{1}{2}} \div a^{\frac{1}{6}} = a^{\frac{1}{2} - \frac{1}{6}} = a^{\frac{3}{6} - \frac{1}{6}} = a^{\frac{2}{6}} = a^{\frac{1}{3}}$.

For $a^{\frac{1}{2}} = a^{\frac{3}{6}} = a^{\frac{1}{6}} \times a^{\frac{1}{6}} \times a^{\frac{1}{6}}$ and this divided by $a^{\frac{1}{6}}$ is

$$\frac{a^{\frac{1}{6}} \times a^{\frac{1}{6}} \times a^{\frac{1}{6}}}{a^{\frac{1}{6}}} = a^{\frac{1}{6}} \times a^{\frac{1}{6}} = a^{\frac{2}{6}} = a^{\frac{1}{3}}.$$

In the same manner, it may be shown that $a^{\frac{1}{m}} \div a^{\frac{1}{n}} = a^{\frac{1}{m} - \frac{1}{n}}$.

Divide	$(3a)^{\frac{11}{12}}$	$(ax)^{\frac{2}{3}}$	$a^{\frac{m+n}{nm}}$	$(b+y)^{\frac{2}{n}}$	$(r^2 y^3)^{\frac{1}{7}}$
By	$a^{\frac{2}{3}}$	$(ax)^{\frac{1}{3}}$	$a^{\frac{1}{m}}$	$(b+y)^{\frac{1}{n}}$	$(r^2 y^3)^{\frac{3}{7}}$
Quot.	$(3a)^{\frac{1}{4}}$		$a^{\frac{1}{n}}$		$(r^2 y^3)^{-\frac{2}{7}}$

Powers and *roots* may be brought promiscuously together, and divided according to the same rule. See art. 281.

Thus $a^2 \div a^{\frac{1}{3}} = a^{2 - \frac{1}{3}} = a^{\frac{5}{3}}$. For $a^{\frac{5}{3}} \times a^{\frac{1}{3}} = a^{\frac{6}{3}} = a^2$.

So $y^n \div y^{\frac{1}{m}} = y^{n - \frac{1}{m}}$.

287. When radical quantities which are reduced to the same index have *rational co-efficients, the rational parts may be divided separately, and their quotient prefixed to the quotient of the radical parts.*

Thus $ac\sqrt{bd} \div a\sqrt{b} = c\sqrt{d}$. For this quotient multiplied into the divisor is equal to the dividend. [Art. 282.]

Divide	$24x\sqrt{ay}$	$18dh\sqrt{bx}$	$by(a^3 x^2)^{\frac{1}{n}}$	$16\sqrt{32}$	$b\sqrt{xy}$
By	$6\sqrt{a}$	$2h\sqrt{x}$	$y(ax)^{\frac{1}{n}}$	$8\sqrt{4}$	$\sqrt{y}$
	$4x\sqrt{y}$		$b(a^2 x)^{\frac{1}{n}}$		$b\sqrt{x}$

Divide $ab(x^2 b)^{\frac{1}{4}}$ by $q(x)^{\frac{1}{2}}$.

These reduced to the same index are $ab(x^2b)^{\frac{1}{4}}$ and $a(x^2)^{\frac{1}{4}}$.

The quotient then is $b(b)^{\frac{1}{4}} = (b^5)^{\frac{1}{4}}$. (Art. 272.)

To save the trouble of reducing to a common index, the division may be expressed in the form of a fraction, as in art. 284.

The quotient will then be $\dfrac{ab(x^2b)^{\frac{1}{4}}}{a(x^{\frac{1}{2}})}$.

INVOLUTION OF RADICAL QUANTITIES.

288. *Radical quantities*, like powers, *are involved by multiplying the index of the root into the index of the required power.*

1. The square of $a^{\frac{1}{3}} = a^{\frac{1}{3} \times 2} = a^{\frac{2}{3}}$. For $a^{\frac{1}{3}} \times a^{\frac{1}{3}} = a^{\frac{2}{3}}$. [Art. 280.]

2. The cube of $a^{\frac{1}{4}} = a^{\frac{1}{4} \times 3} = a^{\frac{3}{4}}$. For $a^{\frac{1}{4}} \times a^{\frac{1}{4}} \times a^{\frac{1}{4}} = a^{\frac{3}{4}}$.

3. And universally, the nth power of $a^{\frac{1}{m}} = a^{\frac{1}{m} \times n} = a^{\frac{n}{m}}$.

For the nth power of $a^{\frac{1}{m}} = a^{\frac{1}{m}} \times a^{\frac{1}{m}} \dots n$ times, and the sum of the indices will then be $\frac{n}{m}$.

4. The 5th power of $a^{\frac{1}{2}} y^{\frac{1}{3}}$, is $a^{\frac{5}{2}} y^{\frac{5}{3}}$. Or, by reducing the roots to a common index,

$$(a^3 y^2)^{\frac{1}{6} \times 5} = (a^3 y^2)^{\frac{5}{6}}.$$

5. The cube of $a^{\frac{1}{n}} x^{\frac{1}{m}}$ is $a^{\frac{3}{n}} x^{\frac{3}{m}}$ or $(a^m x^n)^{\frac{3}{nm}}$.

6. The square of $a^{\frac{2}{3}} x^{\frac{3}{4}}$, is $a^{\frac{4}{3}} x^{\frac{6}{4}}$.

The cube of $a^{\frac{1}{3}}$ is $a^{\frac{1}{3} \times 3} = a^{\frac{3}{3}} = a$. [Art. 253.]

And the nth power of $a^{\frac{1}{n}}$, is $a^{\frac{n}{n}} = a$. That is,

289. *A root is raised to a power of the same name, by removing the index or radical sign.*

Thus the cube of $\sqrt[3]{b+x}$, is $b+x$.

And the nth power of $(a-y)^{\frac{1}{n}}$, is $a-y$.

290. When the radical quantities have *rational co-efficients*, these must also be involved.

1. The square $a\sqrt[n]{x}$, is $a^2 \cdot \sqrt[n]{x^2}$.

For $a\sqrt[n]{x} \times a\sqrt[n]{x} = a^2 \sqrt[n]{x^2}$. [Art. 282.]

2. The nth power of $a^m x^{\frac{1}{m}}$, is $a^{nm} x^{\frac{n}{m}}$.

3. The square of $a\sqrt[3]{x-y}$, is $a^2 \times (x-y)$.

4. The cube of $3a\sqrt[3]{y}$, is $27a^3 y$.

291. But if the radical quantities are connected with others by the signs $+$ and $-$, they must be involved by a multiplication of the several terms, as in art. 213.

Ex. 1. Required the cube of $a+\sqrt{y}$.

$$a+\sqrt{y}$$
$$a+\sqrt{y}$$

$$\overline{\quad\quad}$$

$$a^2 + a\sqrt{y} \quad \text{[Art. 244.]}$$
$$\quad\quad a\sqrt{y}+y \quad \text{[Art. 289.]}$$

$$\overline{\quad\quad}$$

$$a^2 + 2a\sqrt{y}+y$$
$$a + \sqrt{y}$$

$$\overline{\quad\quad}$$

$$a^3 + 2a^2\sqrt{y}+ay$$
$$\quad\quad a^2\sqrt{y}+2ay+y\sqrt{y}$$

$$\overline{\quad\quad}$$

$$a^3 + 3a^2\sqrt{y}+3ay+y\sqrt{y}$$

2. Required the square of $a-\sqrt{b}$. Ans. $a^2 - 2a\sqrt{b}+b$.

3. Required the cube of $2d+\sqrt{x}$.

292. It is unnecessary to give a separate rule for the *evolution* of radical quantities, that is, for finding the root of a quantity which is already a root. The operation is the same as in other cases of evolution. The fractional index of the radical quantity is to be divided, by the number expressing the root to be found. Or, the radical sign belonging to the required root, may be placed over the given quantity. [Art. 257.] If there are rational co-efficients, the roots of these must also be extracted.

Thus, the square root of $a^{\frac{1}{3}}$, is $a^{\frac{1}{3} \div 2} = a^{\frac{1}{6}}$.

For $a^{\frac{1}{6}} \times a^{\frac{1}{6}} = a^{\frac{1}{3}}$.

The cube root of $a(xy)^{\frac{1}{2}}$, is $a^{\frac{1}{3}}(xy)^{\frac{1}{6}}$.

The nth root of $a^{6}\sqrt{by}$, is $\sqrt[n]{a^{6}\sqrt{by}}$.

293. It may be proper to observe, that dividing the *fractional* index of a root is the same in effect, as *multiplying* the number which is placed over the radical sign. For this number corresponds with the *denominator* of the fractional index; and a fraction is divided, by *multiplying* its denominator. [Art. 163.]

$$\text{Thus}^2 \sqrt{a} = a^{\frac{1}{2}}. \qquad\qquad {}^6\sqrt{a} = a^{\frac{1}{6}}.$$

$$^4\sqrt{a} = a^{\frac{1}{4}}. \qquad\qquad {}^{2n}\sqrt{a} = a^{\frac{1}{2n}}.$$

On the other hand, *multiplying* the fractional index is equivalent to *dividing* the number which is placed over the radical sign.

Thus the square of $^6\sqrt{a}$ or $a^{\frac{1}{6}}$, is $^3\sqrt{a}$ or $a^{\frac{1}{6} \times 3} = a^{\frac{1}{3}}$.

REDUCTION OF EQUATIONS BY INVOLUTION AND EVOLUTION.

ART. 294. IN an equation, the letter which expresses the unknown quantity is sometimes found under a *radical sign*. We may have

$$\sqrt{x} = a.$$

To clear this of the radical sign, let each member of the equation be squared, that is, multiplied into itself. We shall then have

$$\sqrt{x} \times \sqrt{x} = aa, \qquad \text{Or, [Art. 289.]} \quad x = a^2.$$

The equality of the sides is not affected by this operation, because each is only multiplied into itself, that is, equal quantities are multiplied into equal quantities. [Ax. 3.]

The same principle is applicable to any root whatever. If $^n\sqrt{x} = a$; then $x = a^n$. For by art. 289, a root is raised to a power of the same name, by removing the index or radical sign. Hence,

295. *When the unknown quantity is under a* RADICAL SIGN, *the equation is reduced by* INVOLVING *both sides*, to a power of the same name, as the root expressed by the radical sign.

It will generally be expedient to make the necessary transpositions, *before* involving the quantities; so that all those which are not under the radical sign, may stand on one side of the equation.

Ex. 1. Reduce the equation $\qquad\quad \sqrt{x} + 4 = 9$
    Transposing $+4$ [Art. 173.] $\quad \sqrt{x} = 9 - 4 = 5$
    Involving both sides $\qquad\qquad x = 5^2 = 25.$

  2. Reduce the equation $\qquad\qquad a + {^n\sqrt{x}} - b = d$
    By transposition, $\qquad\qquad\quad {^n\sqrt{x}} = d + b - a$
    By involution, $\qquad\qquad\qquad x = (d + b - a)^n.$

3. Reduce the equation $\qquad$ $\sqrt[3]{x+1}=4$
Involving both sides, $\qquad$ $x+1=4^3=64$
Transposing $+1$, $\qquad$ $x=65.$

4. Reduce the equation $\qquad$ $4+3\sqrt{x-4}=6+\frac{1}{2}$
Clearing of fractions, [Art.183.] $\quad 8+6\sqrt{x-4}=13$
Transposing 8, $\qquad$ $6\sqrt{x-4}=13-8=5$
Dividing by 6, [Art. 184.] $\qquad$ $\sqrt{x-4}=\frac{5}{6}$
Involving both sides, $\qquad$ $x-4=\frac{25}{36}$ [Art. 223.]
Transposing -4, $\qquad$ $x=\frac{25}{36}+4.$

5. Reduce the equation $\qquad$ $\sqrt{a^2+\sqrt{x}}=\dfrac{3+d}{\sqrt{a^2+\sqrt{x}}}$

Multiplying by $\sqrt{a^2+\sqrt{x}}$, $\quad a^2+\sqrt{x}=3+d$
Transposing a^2, $\qquad$ $\sqrt{x}=3+d-a^2$
Involving both sides, $\qquad$ $x=(3+d-a^2)^2.$

In the first step in this example, multiplying the first member into $\sqrt{a^2+\sqrt{x}}$, that is, into itself, is the same as squaring
it, which is done by taking away its radical sign. The oth-
er member being a fraction, is multiplied into a quantity
equal to its denominator, by cancelling the denominator.
(Art. 159.) There remains a radical sign over x, which
must be removed by involving both sides of the equation.

6. Reduce $3+2\sqrt{x}-\frac{4}{5}=6.$ Ans. $x=\frac{361}{100}.$

7. Reduce $4\sqrt{\dfrac{x}{5}}=8.$ Ans. $x=20.$

REDUCTION OF EQUATIONS BY EVOLUTION.

296. In many equations, the letter which expresses the
unknown quantity is involved to some power. Thus in the
equation

$$x^2=16$$

we have the value of the *square* of x, but not of x itself.
If the square root of both sides be extracted, we shall have

$$x=4.$$

The equality of the members is not affected by this re-
T

duction. For if two quantities or sets of quantities are equal, their roots are also equal.

If $(x+a)^n = b+h$, then $x+a = \sqrt[n]{b+h}$. Hence,

297. *When the expression containing the unknown quantity is a* POWER, *the equation is reduced by* EXTRACTING *the* ROOT *of both sides,* a root of the same name as the power.

Ex. 1. Reduce the equation $\quad\quad 6+x^2-8=7$
By transposition $\quad\quad\quad\quad x^2=7+8-6=9$
By evolution $\quad\quad\quad\quad x=\pm\sqrt{9}=\pm3.$

The signs $+$ and $-$ are both placed before $\sqrt{9}$, because an even root of an affirmative quantity is *ambiguous.* [Art. 261.]

2. Reduce the equation $\quad\quad 5x^2-30=x^2+34$
Trans. and uniting terms, $\quad\quad 4x^2=64$
Dividing by 4, $\quad\quad\quad\quad x^2=16$
By evolution, $\quad\quad\quad\quad x=\pm4.$

3. Reduce the equation $\quad a+\dfrac{x^2}{b}=h-\dfrac{x^2}{d}$

Clearing of fractions, $\quad abd+dx^2-bdh-bx^2$
Transposing terms, $\quad bx^2+dx^2=bdh-abd$

Dividing by $b+d$ (Art.185.) $\quad x^2=\dfrac{bdh-abd}{b+d}$

By evolution $\quad\quad\quad x=\pm\left(\dfrac{bdh-abd}{b+d}\right)^{\frac{1}{2}}$

4. Reduce the equation $\quad a+dx^n=10-x^n$
By transposition, $\quad\quad dx^n+x^n=10-a$

Dividing by $d+1$, $\quad\quad x^n=\dfrac{10-a}{d+1}$

By evolution, $\quad\quad\quad x=\left(\dfrac{10-a}{d+1}\right)^{\frac{1}{n}}.$

298. From the preceding articles, it will be easy to see in what manner an equation is to be reduced, when the expression containing the unknown quantity is a power, and at the same time under a radical sign; that is, when it is a root

of a power. Both involution and evolution will be necessary in this case.

Ex. 1. Reduce the equation $\qquad \sqrt[3]{x^2}=4$
By involution, $\qquad x^2=4^3=64$
By evolution, $\qquad x=\pm\sqrt{64}=\pm8.$

2. Reduce the equation $\qquad \sqrt{x^m-a}=h-d$
By involution $\qquad x^m-a=h^2-2hd+d^2$
Transposing a $\qquad x^m=h^2-2hd+d^2+a$
By evolution $\qquad x=\sqrt[m]{h^2-2hd+d^2+a}.$

3. Reduce the equation $\qquad (x+a)^{\frac{1}{2}}=\dfrac{a+b}{(x-a)^{\frac{1}{2}}}$

Multiplying by $(x-a)^{\frac{1}{2}}$ [Art. 279.] $\quad (x^2-a^2)^{\frac{1}{2}}=a+b$
By involution, $\qquad x^2-a^2=a^2+2ab+b^2$
Trans. and uniting terms, $\qquad x^2=2a^2+2ab+b^2$

By evolution $\qquad x=\pm(2a^2+2ab+b^2)^{\frac{1}{2}}.$

PROBLEMS.

Prob. 1. A gentleman being asked his age, replied; "If you add to it ten years, and extract the square root of the sum, and from this root subtract 2, the remainder will be 6." What was his age?

By the conditions of the problem $\quad \sqrt{x+10}-2=6$
By transposition, $\qquad \sqrt{x+10}=6+2=8$
By involution, $\qquad x+10=8^2=64$
By transposition, $\qquad x=64-10=54.$

Proof [Art. 194.] $\quad \sqrt{54+10}-2=6.$

Prob. 2. If to a certain number 22577 be added, and the square root of the sum be extracted, and from this 163 be subtracted, the remainder will be 237. What is the number?

Let $x=$ the number sought. $\qquad b=163$
$\qquad a=22577$ $\qquad\qquad c=237.$

By the conditions proposed $\sqrt{x+a}-b=c$

By transposition, $\sqrt{x+a}=c+b$

By involution, $x+a=(c+b)^2$

By transposition, $x=(c+b)^2-a$

Restoring the numbers, (Art.52.) $x=(237+163)^2-22577$

That is, $x=160000-22577=137423.$

        Proof $\sqrt{137423+22577}-163=237.$

299. When an equation is reduced by extracting an even root of a quantity, the solution does not determine whether the answer is positive or negative. [Art. 297.] But what is thus left ambiguous by the algebraic process, is frequently settled by the statement of the problem.

Prob. 3. A merchant gains in trade a sum, to which 320 dollars bears the same proportion, as five times this sum does to 2500. What is the amount gained?

          Let $x=$ the sum required.
            $a=320$
            $b=2500.$

By the supposition $a:x::5x:b$

Mult. extremes and means [Art. 188.] $5x^2=ab$

Dividing by 5, $x^2=\dfrac{ab}{5}$

By evolution, $x=\left(\dfrac{ab}{5}\right)^{\frac{1}{2}}$

Restoring the numbers, $x=\left(\dfrac{320\times2500}{5}\right)^{\frac{1}{2}}=400.$

Here the answer is not marked as ambiguous, because by the statement of the problem it is *gain*, and not loss. It must therefore be positive. This might be determined, in the present instance, even from the algebraic process. Whenever the root of x^2 is ambiguous, it is because we are ignorant whether the power has been produced by the multiplication of $+x$, or of $-x$, into itself. [Art. 262.] But here we have the multiplication actually performed. By turning back to the two first steps of the equation, we find that $5x^2$ was produced by multiplying $5x$ into x, that is $+5x$ into $+x$.

Prob. 4. The distance to a certain place is such, that if 96 be subtracted from the square of the number of miles, the remainder will be 48. What is the distance?

Let $x=$ the distance required.

By the supposition,	$x^2 - 96 = 48$
By transposition,	$x^2 = 48 + 96 = 144$
By evolution,	$x = \sqrt{144} = 12.$

Prob. 5. If three times the square of a certain number be divided by four, and if the quotient be diminished by 12, the remainder will be 180. What is the number?

By the supposition	$\dfrac{3x^2}{4} - 12 = 180$
Multiplying by 4, and trans.	$3x^2 = 720 + 48 = 768$
Dividing by 3,	$x^2 = 256$
By evolution,	$x = \sqrt{256} = 16.$

Prob. 6. What number is that, the fourth part of whose square being subtracted from 8, leaves a remainder equal to four? Ans. 4.

AFFECTED QUADRATIC EQUATIONS.

300. Equations are divided into classes, which are distinguished from each other, by the power of the letter that expresses the unknown quantity. Those which contain only the *first* power of the unknown quantity, are called equations of *one dimension*, or equations of the *first degree*. Those in which the highest power of the unknown quantity is a *square*, are called *quadratic*, or equations of the *second degree;* those in which the highest power is a *cube*, equations of the *third degree*, &c.

Thus $x = a + b$, is an equation of the *first* degree.

$x^2 = c$, and $x^2 + ax = d$, are *quadratic* equations, or equations of the *second* degree.

$x^3 = h$, and $x^3 + ax^2 + bx = d$, are *cubic* equations, or equations of the *third* degree.

301. Equations are also divided into *pure* and *affected*

equations. A pure equation contains only *one power* of the unknown quantity. This may be the first, second, third, or any other power. An affected equation contains *different powers* of the unknown quantity. Thus,

$\begin{cases} x^2 = d - b, \text{ is a pure quadratic equation.} \\ x^2 + bx = d, \text{ an affected quadratic equation.} \end{cases}$

$\begin{cases} x^3 = b - c, \text{ a pure cubic equation.} \\ x^3 + ax^2 + bx = h, \text{ an affected cubic equation.} \end{cases}$

A pure equation is also called a *simple* equation. But this term has been applied in too vague a manner. By some writers, it is extended to pure equations of every degree: by others, it is confined to those of the first degree.

In a *pure* equation, all the terms which contain the unknown quantity may be united in one, (Art. 185.) and the equation, however complicated in other respects, may be reduced by the rules which have already been given. But in an *affected* equation, as the unknown quantity is raised to *different powers*, the terms containing these powers can not be united. (Art. 230.) There are particular rules for the reduction of quadratic, cubic, and biquadratic equations. Of these, only the first will be considered at present.

302. *An affected quadratic equation is one which contains the unknown quantity in one term, and the square of that quantity in another term.*

The unknown quantity may be originally in *several* terms of the equation. But all these may be reduced to two, one containing the unknown quantity, and the other its square.

303. It has already been shown that a *pure* quadratic is solved by *extracting the root of both sides of the equation.* An *affected* quadratic may be solved in the same way, if the member which contains the unknown quantity is an *exact square.* Thus the equation

$$x^2 + 2ax + a^2 = b + h$$

may be reduced by evolution. For the first member is the square of a *binomial* quantity. [Art. 264.] And its root is $x + a$. Therefore,

$$x + a = \sqrt{b + h}, \text{ and by transposing } a,$$
$$x = \sqrt{b + h} - a.$$

304. But it is not often the case, that a member of an affected quadratic equation is an exact square, till an addition-

al term is applied, for the purpose of making the required reduction. In the equation

$$x^2 + 2ax = b$$

the side containing the unknown quantity is not a complete square. The two terms of which it is composed are indeed such, as might belong to the square of a binomial quantity. [Art. 214.] But one term is *wanting*. We have then to inquire, in what way this may be supplied. From having *two* terms of the square of a binomial given, how shall we find the *third*?

Of the three terms, two are complete powers, and the other is twice the product of the roots of these powers; [Art. 214.] or, which is the same thing, the product of one of the roots into twice the other. In the expression

$$x^2 + 2ax,$$

the term $2ax$ consists of the factors $2a$ and x. The latter is the unknown quantity. The other factor $2a$ may be considered the *co-efficient* of the unknown quantity; a co-efficient being another name for a factor. [Art. 41.] As x is the root of the first term x^2; the other factor $2a$ is *twice* the root of the third term, which is wanted to complete the square. Therefore *half* $2a$ is the root of the deficient term, and a^2 is the term itself. The square completed is

$$x^2 + 2ax + a^2,$$

where it will be seen that the last term a^2 is the square of half $2a$, and $2a$ is the co-efficient of x the root of the first term.

In the same manner, it may be proved, that the last term of the square of any binomial quantity, is equal to the square of half the co-efficient of the root of the first term. From this principle, is derived the following rule:

305. To *complete the square*, in an affected quadratic equation; *take the square of half the co-efficient of the first power of the unknown quantity, and add it to both sides of the equation.*

Before completing the square, the known and unknown quantities must be brought on opposite sides of the equation by transposition; and the highest power of the unknown quantity must have the affirmative sign, and be cleared of fractions, co-efficients, &c. See arts. 308, 9, 10, 11.

After the square is completed, the equation is reduced, by extracting the square root of both sides, and transposing the known part of the binomial root. [Art. 303.]

The quantity which is added to one side of the equation, to complete the square, must be added to the other side also, to preserve the equality of the two members. (Ax. 1.)

306. To avoid having the attention divided among too many objects, the learner should distinguish between what is *peculiar* in the reduction of quadratic equations, and what is common to this and the other kinds which have already been considered. The peculiar part, in the resolution of affected quadratics, is the *completing of the square*. The other steps are similar to those by which pure equations are reduced.

For the purpose of rendering the completing of the square familiar, there will be an advantage in beginning with examples in which the equation is already prepared for this step.

Ex. 1. Reduce the equation $\qquad$ $x^2 + 6ax = b$

Completing the square $\qquad$ $x^2 + 6ax + 9a^2 = 9a^2 + b$

Extracting both sides (Art.303.) $x + 3a = \pm\sqrt{9a^2 + b}$

Transposing $3a$ $\qquad$ $x = -3a \pm \sqrt{9a^2 + b}$

Here the co-efficient of x, in the first step, is $6a$; The square of half this is $9a^2$, which being added to both sides completes the square. The equation is then reduced by extracting the root of each member, in the same manner as in art. 297, excepting that the square here being that of a *binomial*, its root is found by the rule in art. 265.

2. Reduce the equation $\qquad$ $x^2 - 8bx = h$

Completing the square, $\qquad$ $x^2 - 8bx + 16b^2 = 16b^2 + h$

Extracting both sides $\qquad$ $x - 4b = \pm\sqrt{16b^2 + h}$

Transposing $-4b$, $\qquad$ $x = 4b \pm \sqrt{16b^2 + h}$

In this example, half the co-efficient of x is $4b$, the square of which $16b^2$ is to be added to both sides of the equation.

3. Reduce the equation $\qquad$ $x^2 + ax = b + h$

Completing the square, $\qquad$ $x^2 + ax + \dfrac{a^2}{4} = \dfrac{a^2}{4} + b + h$

By evolution, $\qquad$ $x + \dfrac{a}{2} = \pm\left(\dfrac{a^2}{4} + b + h\right)^{\frac{1}{2}}$

Transposing $\dfrac{a}{2}$ $\qquad$ $x = \dfrac{a}{2} \pm \left(\dfrac{a^2}{4} + b + h\right)^{\frac{1}{2}}.$

Here the co-efficient of x is a, half of which is $\frac{a}{2}$, whose square is $\frac{a^2}{4}$. (Art. 223.)

4. Reduce the equation $\qquad x^2 - x = h - d$

Completing the square, $\qquad x^2 - x + \frac{1}{4} = \frac{1}{4} + h - d$

Extracting and transp. $\qquad x = \frac{1}{2} \pm (\frac{1}{4} + h - d)^{\frac{1}{2}}.$

Here the co-efficient of x is 1, its half is $\frac{1}{2}$, and the square of this is $\frac{1}{4}$.

5. Reduce the equation $\qquad x^2 + 3x = d + 6$

Completing the square, $\qquad x^2 + 3x + \frac{9}{4} = \frac{9}{4} + d + 6$

Extracting and transp. $\qquad x = -\frac{3}{2} \pm (\frac{9}{4} + d + 6)^{\frac{1}{2}}.$

6. Reduce the equation $\qquad x^2 - abx = ab - cd$

Completing the square, $\quad x^2 - abx + \dfrac{a^2 b^2}{4} = \dfrac{a^2 b^2}{4} + ab - cd$

Extracting and transp. $\quad x = \dfrac{ab}{2} \pm \left(\dfrac{a^2 b^2}{4} + ab - cd \right)^{\frac{1}{2}}.$

7. Reduce the equation $\qquad x^2 + \dfrac{ax}{b} = h$

Completing the square, $\quad x^2 + \dfrac{ax}{b} + \dfrac{a^2}{4b^2} = \dfrac{a^2}{4b^2} + h$

Extracting and transp. $\quad x = -\dfrac{a}{2b} \pm \left(\dfrac{a^2}{4b^2} + h \right)^{\frac{1}{2}}.$

By art. 158, $\dfrac{ax}{b} = \dfrac{a}{b} \times x$. The co-efficient of x, therefore, is $\dfrac{a}{b}$. Half of this is $\dfrac{a}{2b}$, (Art. 163.) the square of which is $\dfrac{a^2}{4b^2}$.

U

8. Reduce the equation, $\quad x^2 - \dfrac{x}{b} = 7h$

Completing the square, $\quad x^2 - \dfrac{x}{b} + \dfrac{1}{4b^2} = \dfrac{1}{4b^2} + 7h$

Extracting and transp. $\quad x = \dfrac{1}{2b} - \left(\dfrac{1}{4b^2} + 7h \right)^{\frac{1}{2}}.$

Here the fraction $\dfrac{x}{b} = \dfrac{1}{b} \times x.$ (Art. 158.) Therefore the co-efficient of x is $\dfrac{1}{b}.$

307. In these and similar instances, the root of the third term of the completed square is easily found, because this root is the same half co-efficient from which the term has just been derived. (Art. 304.) Thus in the last example, half the co-efficient of x is $\dfrac{1}{2b}$, and this is the root of the third term $\dfrac{1}{4b^2}.$

308. When the first power of the unknown quantity is in *several terms*, these should be united in one, if they can be, by the rules for reduction in addition. But if there are *literal* co-efficients, these may be considered as constituting, together, a *compound* co-efficient or factor, into which the unknown quantity is multiplied.

Thus $ax + bx + dx = (a + b + d) \times x.$ (Art. 120.) The square of half this compound co-efficient is to be added to both sides of the equation.

1. Reduce the equation, $\quad x^2 + 3x + 2x + x = d$
   Uniting terms, (Art 174.) $\quad x^2 + 6x = d$
   Completing the square, $\quad x^2 + 6x + 9 = 9 + d$
   Extracting and transp. $\quad x = -3 \pm \sqrt{9 + d}$

2. Reduce the equation $\quad x^2 + ax + bx = h$
   By art. 120, $\quad x^2 + (a + b) \times x = h$

Compl'g the square, $x^2 + (a+b) \times x + \left(\dfrac{a+b}{2} \right)^2 = \left(\dfrac{a+b}{2} \right)^2 + h$

By evolution, $\quad x + \dfrac{a+b}{2} = \pm \sqrt{ \left(\dfrac{a+b}{2} \right)^2 + h }$

By transposition, $\quad x = -\dfrac{a+b}{2} \pm \sqrt{ \left(\dfrac{a+b}{2} \right)^2 + h }$

3. Reduce the equation, $x^2 + ax - x = b$

By art. 120, $\qquad x^2 + (a-1) \times x = b$

Compl'g the square, $x^2 + (a-1) \times x + \left(\dfrac{a-1}{2}\right)^2 = \left(\dfrac{a-1}{2}\right)^2 + b$

Extracting and transp. $x = -\dfrac{a-1}{2} \pm \sqrt{\left(\dfrac{a-1}{2}\right)^2 + b}$

309. After becoming familiar with the method of completing the square, in affected quadratic equations, it will be necessary to attend to the steps which are *preparatory* to this. Here however, little more is necessary, than an application of rules already given. The known and unknown quantities must be brought on opposite sides of the equation by transposition. And it will generally be expedient to make the square of the unknown quantity the first or leading term, as in the preceding examples. This indeed is not essential. But it will show, to the best advantage, the arrangement of the terms in the completed square.

1. Reduce the equation $\qquad a + 5x - 3b = 3x - x^2$

Transp. and uniting terms, $\quad x^2 + 2x = 3b - a$

Completing the square, $\qquad x^2 + 2x + 1 = 1 + 3b - a$

Extracting and transp. $\qquad x = -1 \pm \sqrt{1 + 3b - a}.$

2. Reduce the equation $\qquad \dfrac{x}{2} = \dfrac{36}{x+2} - 4$

Clearing of fractions, $\qquad x^2 + 2x = 72 - 8x - 16$

Transp. and uniting terms, $\quad x^2 + 10x = 56$

Completing the square, $\qquad x^2 + 10x + 25 = 25 + 56 = 81$

Extracting and transp. $\qquad x = -5 \pm \sqrt{81} = -5 \pm 9.$

310. If the *highest power* of the unknown quantity has any co-efficient, or *divisor*, it must, *before* the square is completed, be freed from these, by multiplication or division, as in arts. 180 and 184.

1. Reduce the equation $\qquad x^2 + 24a - 6h = 12x - 5x^2$

Transp. and uniting terms, $\quad 6x^2 - 12x = 6h - 24a$

Dividing by 6, $\qquad x^2 - 2x = h - 4a$

Completing the square, $\qquad x^2 - 2x + 1 = 1 + h - 4a$

Extracting and transp. $\qquad x = 1 \pm \sqrt{1 + h - 4a}$

2. Reduce the equation $\quad h+2x=d-\dfrac{bx^2}{a}$

Clearing of fractions, $\quad ah+2ax=ad-bx^2$

By transposition, $\quad bx^2+2ax=ad-ah$

Dividing by b, $\quad x^2+\dfrac{2ax}{b}=\dfrac{ad-ah}{b}$

Compl'g the square, $\quad x^2+\dfrac{2ax}{b}+\dfrac{a^2}{b^2}=\dfrac{a^2}{b^2}+\dfrac{ad-ah}{b}$

Extracting and transp. $\quad x=-\dfrac{a}{b}+\left(\dfrac{a^2}{b^2}+\dfrac{ad-ah}{b}\right)^{\frac{1}{2}}$

311. If the square of the unknown quantity is in *several terms*, the equation must be divided by *all* the co-efficients of this square, as in art. 185.

1. Reduce the equation $\quad bx^2+dx^2-4x=b-h$

Dividing by $b+d$, (Art. 121.)$x^2-\dfrac{4x}{b+d}=\dfrac{b-h}{b+d}$

Completing the square $x^2-\dfrac{4x}{b+d}+\left(\dfrac{2}{b+d}\right)^2=\left(\dfrac{2}{b+d}\right)^2+\dfrac{b-h}{b+d}$

Extract. and transp. $\quad x=\dfrac{2}{b+d}+\sqrt{\left(\dfrac{2}{b+d}\right)^2+\dfrac{b-h}{b+d}}$

2. Reduce the equation $\quad ax^2+x=h+3x-x^2$

Transp. and uniting terms, $\quad ax^2+x^2-2x=h$

Dividing by $a+1$, $\quad x^2-\dfrac{2x}{a+1}=\dfrac{h}{a+1}$

Completing the square, $x^2-\dfrac{2x}{a+1}+\left(\dfrac{1}{a+1}\right)^2=\left(\dfrac{1}{a+1}\right)^2+\dfrac{h}{a+1}$

Extracting and transp. $x=\dfrac{1}{a+1}+\sqrt{\left(\dfrac{1}{a+1}\right)^2+\dfrac{h}{a+1}}$

312. In the square of a binomial, the first and last terms are always *positive*. For each is the square of one of the terms of the root. (Art. 214.) But every square is positive. (Art. 218.) If then $-x^2$ occurs in an equation, it can not, with this sign, form a part of the square of a binomial. But if *all* the signs in the equation be changed, the equality of the sides will be preserved, (Art. 177.) the term $-x^2$ will become positive, and the square may be completed.

1. Reduce the equation $\qquad -x^2+2x=d-h$
Changing all the signs, $\qquad x^2-2x=h-d$
Completing the square, $\qquad x^2-2x+1=1+h-d$
Extracting and transp. $\qquad x=1\pm\sqrt{1+h-d}.$

2. Reduce the equation $\qquad 4x-x^2=-12$
Changing all the signs, $\qquad x^2-4x=12$
Completing the square, $\qquad x^2-4x+4=4+12=16$
Extracting and transp. $\qquad x=2\pm\sqrt{16}.$

313. In a quadratic equation, the first term x^2 is the square of a single letter. But a binomial quantity may consist of terms, one or both of which are already powers.

Thus x^3+a is a binomial, and its square is

$$x^6+2ax^3+a^2,$$

where the index of x in the first term is twice as great as in the second. When the third term is deficient, the square may be completed in the same manner as that of any other binomial. For the middle term is twice the product of the roots of the two others.

So the square of x^n+a, is $x^{2n}+2ax^n+a^2$. Therefore,

314. *Any equation which contains only two different powers of the unknown quantity, the index of one of which is twice that of the other, may be resolved in the same manner as a quadratic equation, by completing the square.*

It must be observed however that, in the binomial root, the letter expressing the unknown quantity will still have an index, so that a farther extraction, according to art. 297, will be necessary.

Reduce the equation $\qquad x^4-x^2=b-a$
Completing the square, $\qquad x^4-x^2+\frac{1}{4}=\frac{1}{4}+b-a$
Extracting and transp. $\qquad x^2=\frac{1}{2}\pm\sqrt{\frac{1}{4}+b-a}$
Extracting again, (Art.297.) $x=\sqrt{\frac{1}{2}\pm\sqrt{\frac{1}{4}+b-a}}.$

2. Reduce the equation $\qquad x^{2n}-4bx^n=a$
Completing the square, $\qquad x^{2n}-4bx^n+4b^2=4b^2+a$
Extracting and transp. $\qquad x^n=2b\pm\sqrt{4b^2+a}$
Extracting again $\qquad x=\sqrt[n]{2b\pm\sqrt{4b^2+a}}.$

315. The solution of a quadratic equation, whether pure

or affected, gives two results. For after the equation is reduced, it contains an ambiguous root. In a *pure* quadratic,
this root is the *whole* value of the unknown quantity. (Art.
297.)

Thus the equation $x^2 = 64$
Becomes, when reduced, $x = \pm \sqrt{64}$

That is, the value of x is either $+8$ or -8, for each
of these is a root of 64. Here both the values of x are the
same, except that they have contrary signs. This will be
the case in every pure quadratic equation, because the whole
of the second member is under the radical sign. The two
values of the unknown quantity will be alike, except that one
will be positive, and the other negative.

316. But in *affected* quadratics, a *part* only of one side of
the reduced equation is under the radical sign. When this
part is added to, or subtracted from, that which is without the
radical sign; the two results will differ in quantity, and will
have their signs in some cases alike, and in others unlike.

1. The equation $x^2 + 8x = 20$
Becomes, when reduced, $x = -4 \pm \sqrt{16 + 20}$
That is, $x = -4 \pm 6.$

Here the first value of x is, $-4 + 6 = +2$ } one positive, and
And the second is, $\quad -4 - 6 = -10$ } the other negative.

2. The equation $x^2 - 8x = -15$
Becomes, when reduced, $x = 4 \pm \sqrt{16 - 15}$
That is $x = 4 \pm 1$

Here the first value of x is $4 + 1 = +5$ } both positive.
And the second is $\quad 4 - 1 = +3$ }

That these two values of x are correctly found, may be
proved, by substituting first one, and then the other, for x itself, in the original equation. (Art. 194.)

Thus $5^2 - 8 \times 5 = 25 - 40 = -15$
And $3^2 - 8 \times 3 = 9 - 24 = -15.$

317. In the reduction of an affected quadratic equation,
the value of the unknown quantity is frequently found to be
imaginary.

Thus the equation $\qquad x^2-8x=-20$

Becomes, when reduced, $\qquad x=4\pm\sqrt{16-20}$

That is, $\qquad x=\pm\sqrt{-4}.$

Here the root of the negative quantity -4 can not be assigned, (Art. 263.) and therefore the value of x can not be found. There will be the same impossibility, in every instance in which the negative part of the quantities under the radical sign is greater than the positive part.[*]

318. Whenever *one* of the values of the unknown quantity, in a quadratic equation is imaginary, the *other* is so also. For both are equally affected by the imaginary root.

Thus, in the example above,

The first value of x is $\qquad 4+\sqrt{-4},$

And the second is $\qquad 4-\sqrt{-4};$ each of which

contains the imaginary quantity $\qquad \sqrt{-4}.$

319. An equation which when reduced contains an imaginary root, is often of use, to enable us to determine whether a proposed question admits of an answer, or involves an absurdity.

Suppose it is required to divide 8 into two such parts, that the product will be 20.

If x is one of the parts, the other will be $8-x$. (Art. 195.)

By the conditions proposed, $\qquad (8-x)\times x=20$

That is, $\qquad 8x-x^2=20$

Changing all the signs, (Art. 177.) $\qquad x^2-8x=-20$

This becomes, when reduced, $\qquad x=4-\sqrt{-4}.$

Here the imaginary expression $\sqrt{-4}$ shows that an answer is impossible; and that there is an absurdity in supposing that 8 may be divided into two such parts, that their product shall be 20.

320. Although a quadratic equation has two solutions, yet both these may not always be applicable to the subject proposed. The quantity under the radical sign may be produced either from a positive or a negative root. But both these roots may not, in every instance, belong to the problem to be solved. See art. 299.

[*] See note C.

Problems producing Quadratic Equations.

Prob. 1. A merchant has a piece of cotton cloth, and a piece of silk. The number of yards in both is 110; and if the square of the number of yards of silk be subtracted from 80 times the number of yards of cotton, the difference will be 400. How many yards are there in each piece?

$$\text{Let } x = \text{the yards of silk.}$$
$$\text{Then } 110 - x = \text{the yards of cotton.}$$

By supposition, $\quad 400 = 80 \times (110 - x) - x^2$
That is, $\quad\quad\quad 400 = 8800 - 80x - x^2$
Transp. & unit. terms, $x^2 + 80x = 8400$
Compl'g the square, $x^2 + 80x + 1600 = 1600 + 8400 = 10000$
Extracting and transp. $x = -40 \pm \sqrt{10000} = -40 \pm 100$

The first value of x, is $-40 + 100 = 60$, the yards of silk;
And $\quad\quad\quad 110 - x = 110 - 60 = 50$, the yards of cotton.
The second value of x, is $-40 - 100 = -140$; but as this is a negative quantity, it is not applicable to goods which a man has in his possession.

Prob. 2. The ages of two brothers are such, that their sum is 45 years, and their product 500. What is the age of each?

$$\text{Let } x = \text{one of the ages.} \quad \text{Then } 45 - x = \text{the other.}$$

By supposition, $\quad x \times (45 - x) = 500$
That is, $\quad\quad\quad 45x - x^2 = 500$
Changing all the signs, $x^2 - 45x = -500$
Compl'g the square, $\quad x^2 - 45x + \dfrac{2025}{4} = \dfrac{2025}{4} - 500 = \dfrac{25}{4}$

Extract. and transp. $\quad x = \dfrac{45}{2} + \sqrt{\dfrac{25}{4}} = \dfrac{45}{2} \pm \dfrac{5}{2}$

One of the ages then is 25 years, and the other 20.

Prob. 3. To find two numbers such, that their difference shall be 4, and their product 117.

Let $x =$ one number, and $x + 4 =$ the other.

By the conditions, $(x+4) \times x = 117$

This reduced, gives, $x = -2 \pm \sqrt{121} = -2 \pm 11.$

One of the numbers therefore is 9, and the other 13.

Prob. 4. A merchant having sold a piece of cloth which cost him 30 dollars, found that if the price for which he sold it were multiplied by his *gain*, the product would be equal to the cube of his gain. What was his gain?

Let $x =$ the gain.

Then $30 + x =$ the price for which the cloth was sold.

By the statement, $x^3 = (30 + x) \times x$

That is, $x^3 = 30x + x^2$

Dividing by x, (Art. 186.) $x^2 = 30 + x$

Transposing x; $x^2 - x = 30$

Completing the square, $x^2 - x + \frac{1}{4} = \frac{1}{4} + 30$

Extracting and transposing $x = \frac{1}{2} \pm \sqrt{\frac{1}{4} + 30} = \frac{1}{2} \pm \frac{11}{2}$

The first value of x is $\frac{1}{2} + \frac{11}{2} = +6.$ }

The second value is $\frac{1}{2} - \frac{11}{2} = -5.$ }

As the last answer is *negative*, it is to be rejected as inconsistent with the nature of the problem, (Art. 320.) for *gain* must be considered *positive*.

Prob. 5. To find two numbers, whose difference shall be 3, and the difference of their cubes 117.

Let $x =$ the least number.

Then $x + 3 =$ the greatest.

By supposition, $(x+3)^3 - x^3 = 117$

Expanding $(x+3)^3$ (Art.217.) $9x^2 + 27x = 117 - 27 = 90$

Dividing by 9, $x^2 + 3x = 10$

Completing the square, $x^2 + 3x + \frac{9}{4} = \frac{9}{4} + 10 = \frac{49}{4}$

Extracting and transp. $x = -\frac{3}{2} \pm \sqrt{\frac{49}{4}} = -\frac{3}{2} \pm \frac{7}{2}.$

The two numbers, therefore, are 2 and 5.

Prob. 6. To find two numbers, whose difference shall be 12, and the sum of their squares 1424.

Ans. The numbers are 20 and 32.

V

Prob. 7. Two persons draw prizes in a lottery, the difference of which is 120 dollars, and the greater is to the less, as the less to 10. What are the prizes?

Let $x =$ the less prize.
Then $v + 120 =$ the greater.

By the statement,	$x + 120 : x :: x : 10$
Mult. extremes and means,	$x^2 = 10x + 1200$
Transposing $10x$,	$x^2 - 10x = 1200$
Completing the square,	$x^2 - 10x + 25 = 25 + 1200$
Extracting and transp.	$x = 5 + \sqrt{25 + 1200} = 5 + 35.$

The two prizes, then, are 40 and 160.

Prob. 8. What two numbers are those whose sum is 6, and the sum of their cubes 72?

Ans. 2 and 4.

SUBSTITUTION.

321. In the reduction of Quadratic Equations, as well as in other parts of algebra, a complicated process may be rendered much more simple, by introducing a new letter which shall be made to represent several others. This is termed *substitution*. A letter may be put for a compound quantity as well as for a single number. Thus in the equation

$$x^2 - 2ax = \tfrac{3}{4} + \sqrt{86 - 64} + h,$$

we may substitute b, for $\tfrac{3}{4} + \sqrt{86 - 64} + h$. The equation will then become $x^2 - 2ax = b$,   and when reduced will be $x = a \pm \sqrt{a^2 + b}.$

After the operation is completed, the compound quantity for which a single letter has been substituted, may be *restored*. The last equation, by restoring the value of b, will become

$$x = a \pm \sqrt{a^2 + \tfrac{3}{4} + \sqrt{86 - 64} + h}.$$

Reduce the equation $\qquad ax-2x-d=bx-x^2-x$

Transp. and uniting terms, $x^2+ax-bx-x=d$

By art. 120, $\qquad x^2+(a-b-1)\times x=d$

Substituting h for $(a-b-1)$, $x^2+hx=d$

Completing the square, $\qquad x^2+hx+\dfrac{h^2}{4}=\dfrac{h^2}{4}+d$

Extracting and transp. $\qquad x=-\dfrac{h}{2}\overset{+}{\underset{-}{}}\sqrt{\dfrac{h^2}{4}+d}$

Restoring the value of h, $\quad x=-\dfrac{a-b-1}{2}\overset{+}{\underset{-}{}}\sqrt{\dfrac{(a-b-1)^2}{4}+d}$

SOLUTION OF PROBLEMS WHICH CONTAIN TWO OR MORE UNKNOWN QUANTITIES. DEMONSTRATION OF THEOREMS.

ART. 322. IN the examples which have been given of the resolution of equations, in the preceding sections, each problem has contained only *one* unknown quantity. Or if, in some instances, there have been *two*, they have been so related to each other, that both have been expressed by means of the same letter. (Art. 195.)

But cases frequently occur in which *several* unknown qaantities are introduced into the same calculation. And if the problem is of such a nature, as to admit of a determinate answer, there will arise from the conditions, as many equations independent of each other, as there are unknown quantities.

Equations are said to be *independent*, when they express different conditions; and *dependent*, when they express the same conditions under different forms. The former are not convertible into each other. But the latter may be changed from one form to the other, by the methods of reduction which have been considered. Thus $b-x=y$, and $b=y+x$, are dependent equations, because one is formed from the other by merely transposing x.

323. In solving a problem, it is necessary first to find the value of one of the unknown quantities, and then of the others in succession. To do this, we must derive from the equations which are given, a new equation, from which all the unknown quantities except one shall be excluded.

Suppose the following equations are given.

$$1. \; x+y=14$$
$$2. \; x-y=2.$$

If y be transposed in each, they will become

$$1. \; x=14-y$$
$$2. \; x=2+y.$$

Here the first member of each of the equations is x, and the second member of each is *equal* to x. But according to axiom 5th, quantities which are respectively equal to any other quantity are equal to each other; therefore,

$$2+y=14-y.$$

Here we have a new equation, which contains only the unknown quantity y.

Transposing 2 and $-y$, $\qquad 2y=12$
Dividing by 2, $\qquad\qquad\quad y=6.$

The value of y is therefore found. Hence,

324. RULE I. To exterminate one of two unknown quantities, and deduce one equation from two; *Find the value of one of the unknown quantities in each of the equations, and form a new equation by making one of these values equal to the other.*

That quantity which is the least involved should be the one which is chosen to be exterminated.

For the convenience of referring to different parts of a solution, the several steps will, in future, be numbered. When an equation is formed from one *immediately preceding*, it will be unnecessary to specify it. In other cases, the number of the equation or equations from which a new one is derived will be referred to.

Prob. 1. To find two numbers such, that
   Their sum shall be 24; and
   The greater shall be equal to 5 times the less.

  Let $x=$the greater;   And $y=$the less.

1. By the first condition, $x+y=24$ ⎫
2. By the second, $x=5y$ ⎬
3. Transp. y in the 1st equation, $x=24-y$
4. Making the 2d and 3d equal, $5y=24-y$
5. Transp. and uniting terms, $6y=24$
6. Dividing by 6, $y=4$, the less number.

Prob. 2. To find one of two quantities,
   Whose sum is equal to h; and
   The difference of whose squares is equal to d.

  Let $x=$the greater quantity;   And $y=$the less.

1. By the first condition,　　　　　　$x+y=h$ ⎫
2. By the second,　　　　　　　　　$x^2-y^2=d$ ⎬ .
3. Transp. y^2 in the 2d equation,　$x^2=d+y^2$　　　⎭
4. By evolution, (Art. 297.)　　　$x=\sqrt{d+y^2}$
5. Transp. y in the 1st equation,　$x=h-y$
6. Making the 4th and 5th equal　$\sqrt{d+y^2}=h-y$
7. By involution, (Art. 295.)　　$d+y^2=h^2-2hy+y^2$
8. Expunging y^2 (Art. 176.)　　$d=h^2-2hy$
9. By transposition,　　　　　　$2hy=h^2-d$

10. Dividing by $2h$,　　　　　　$y=\dfrac{h^2-d}{2h}$.

Prob. 3. Given $ax+by=h$ ⎫ To find y. Ans. $y=\dfrac{h-ad}{b-a}$.
　　　　And　　$x+y=d$ ⎭

325. The rule given above may be generally applied, for the extermination of unknown quantities. But there are cases, in which other methods will be found more expeditious.

Suppose $x=hy$ ⎫
And　$ax+bx=y^2$ ⎬

As in the first of these equations x is equal to hy, we may, in the second equation, *substitute* this value of x instead of x itself. The second equation will then be converted into

$$ahy+bhy=y^2.$$

The equality of the two sides is not affected by this alteration, because we only exchange one quantity x, for another which is equal to it. By this means we obtain an equation which contains only one unknown quantity. Hence,

326. Rule II. To exterminate an unknown quantity, *Find the value of one of the unknown quantities, in one of the equations;* and then, in the other *equation,* substitute *this value, for the unknown quantity itself.*

Prob. 4. A privateer in chase of a ship 20 miles distant, sails 8 miles, while the ship sails 7. How far must the privateer sail, before she overtakes the ship?

It is evident that the whole distance which the privateer sails during the chase, must be to the distance which the ship sails in the same time, as 8 to 7.

Let x=the distance which the privateer sails;
And y=the distance which the ship sails.

1. By the supposition, $x=y+20$ ⎱
2. And also, $x:y::8:7$ ⎰
3. Mult. extremes and means, $8y=7x$
4. Dividing by 8, $y=\frac{7}{8}x$
5. Substituting $\frac{7}{8}x$ for y in the 1st eq. $x=\frac{7}{8}x+20$
6. Multiplying by 8, and transp. $x=160$.

Prob. 5. The ages of two persons A and B are such, that seven years ago, A was three times as old as B; and seven years hence, A will be twice as old as B. What is the age of B?

Let x=the age of A;   And y=the age of B;
Then $x-7$ was the age of A, 7 years ago;
And $y-7$ was the age of B, 7 years ago.
Also $x+7$ will be the age of A, 7 years hence;
And $y+7$ will be the age of B, 7 years hence.

1. By the first condition, $x-7=3\times(y-7)=3y-21$ ⎱
2. By the second, $x+7=2\times(y+7)=2y+14$ ⎰
3. Transp. 7 in the 1st equa. $x=3y-14$
4. Subst. $3y-14$ for x, in the 2d, $3y-14+7=2y+14$
5. Transp. and uniting terms, $y=21$, the age of B.

Prob. 6. There are two numbers, of which
     The greater is to the less, as 3 to 2; and
     Their sum is the sixth part of their product.
       What is the less number?   Ans. 10.

327. There is a *third* method of exterminating an unknown quantity from an equation, which, in many cases, is preferable to either of the preceding.

     Suppose that $x+3y=a$ ⎱
     And that $x-3y=b$ ⎰

If we *add together* the first members of these two equations, and also the second members, we shall have

$$2x=a+b$$

an equation which contains only the unknown quantity x. The other, having equal co-efficients with contrary signs, has disappeared. (Art. 77.) The equality of the sides is preserved, because we have only added equal quantities to equal quantities. (Ax. 1.)

$$\text{Again suppose } \left.\begin{array}{l} 3x+y=h \\ 2x+y=d \end{array}\right\}$$

And

If we *subtract* the last equation from the first, we shall have

$$x=h-d$$

where y is exterminated, without affecting the equality of the sides. (Ax. 2.)

$$\text{Again, suppose } \left.\begin{array}{l} x-2y=a \\ x+4y=b \end{array}\right\}$$

And

Multiplying the 1st by 2, $2x-4y=2a$
Then adding the 2d and 3d, $3x=b+2a$. Hence,

328. RULE III. To exterminate an unknown quantity,

MULTIPLY or DIVIDE *the equation, if necessary, in such a manner that the term which contains one of the unknown quanties shall be the same in both.*

Then SUBTRACT *one equation from the other, if the signs of this unknown quantity are* ALIKE, *or* ADD *them together, if the signs are* UNLIKE.

It must be kept in mind that both members of an equation are always to be increased or diminished, multiplied or divided alike. (Art. 170.)

Prob. 7. The numbers in two opposing armies are such, that,

The sum of both is 21110; and

Twice the number in the greater army, added to three times the number in the less, is 52219.

What is the number in the greater army?

Let x =the greater. And y =the less.

1. By the first condition, $\left.\begin{array}{l} x+y=21110 \\ 2x+3y=52219 \end{array}\right\}$
2. By the second,
3. Multiplying the 1st by 3, $3x+3y=63330$
4. Subtracting the 2d from the 3d, $x=11111.$

Prob. 8. Given $2x+y=16$, and $3x-3y=6$, to find the value of x.

1. By supposition, $2x+y=16$)
2. And $3x-3y=6$)
3. Multiplying the 1st by 3, $6x+3y=48$
4. Adding the 2d and 3d, $9x=54$
5. Dividing by 9, $x=6.$

Prob. 9. Given $x+y=14$, and $x-y=2$, to find the value of y.   Ans. 6.

In the succeeding problems, either of the three rules for exterminating unknown quantities will be made use of, as will in each case be most convenient.

329. When *one* of the unknown quantities is determined, the other may be easily obtained, by going back to an equation which contains both, and substituting, instead of that which is already found, its numerical value.

Prob. 10. The mast of a ship consists of two parts:
One third of the lower part, added to one sixth of the upper part, is equal to 28; and
Five times the lower part, diminished by six times the upper part, is equal to 12.
What is the height of the mast?

Let $x=$the lower part;   And $y=$the upper part.

1. By the first condition, $\frac{1}{3}x+\frac{1}{6}y=28$)
2. By the second, $5x-6y=12$)
3. Multiplying the 1st by 6, $2x+y=168$
4. Dividing the 2d by 6, $\frac{5}{6}x-y=2$
5. Adding the 3d and 4th, $2x+\frac{5}{6}x=170$
6. Multiplying by 6, $12x+5x=1020$
7. Uniting terms, and dividing by 17, $x=60$, the lower part.

Then by the 3d step, $2x+y=168$
That is, substituting 60 for x, $120+y=168$ [per part.
Transposing 120, $y=168-120=48$, the up-

Prob. 11. To find a fraction such that,
If a unit be added to the numerator, the fraction will be equal to $\frac{1}{3}$; but
If a unit be added to the denominator, the fraction will be equal to $\frac{1}{4}$.

Let $x=$the numerator,   And $y=$the denominator.

W

1. By the first condition, $\dfrac{x+1}{y}=\frac{1}{3}$

2. By the second, $\dfrac{x}{y+1}=\frac{1}{4}$

3. Clearing the 1st of fractions, $3x+3=y$

4. Subst. $3x+3$, for y in the 2d, $\dfrac{x}{3x+4}=\frac{1}{4}$

5. Clearing of fractions, $4x=3x+4$

6. Transp. and uniting terms, $x=4$, the numerator

Then subst. 4 for x in the 3d, $12+3=15=y$, the denominator.

Prob. 12. What two numbers are those,
　　　Whose *difference* is to their sum, as 2 to 3; and
　　　Whose sum is to their product, as 3 to 5?
　　　　　　　　　　　　　　　Ans. 10 and 2.

Prob. 13. To find two numbers such, that
　　　The product of their sum and difference shall be 5, and
　　　The product of the sum of their squares and the differ-
　　　ence of their squares shall be 65.

　Let $x=$ the greater number;　　　And $y=$ the less.

1. By the first condition, 　$(x+y)\times(x-y)=5$
2. By the second, 　　$(x^2+y^2)\times(x^2-y^2)=65$
3. Mult. the factors in the 1st, (Art. 235.) $x^2-y^2=5$
4. Dividing the 2d by the 3d, (Art. 118.) $x^2+y^2=13$
5. Adding the 3d and 4th, 　　　$2x^2=18$
6. Dividing by 2, 　　　　　$x^2=9$
7. By evolution, 　　$x=\sqrt{9}=3$, the greater number.
　The other number is 　2.

In the 4th step, the first member of the 2d equation is divided by x^2-y^2, and the second member by 5, which is
equal to x^2-y^2.

Prob. 14. To find two numbers, whose difference is 8, and
product 240.

Prob. 15. To find two numbers,
　　　Whose difference shall be 12, and
　　　The sum of their squares 1424.

　Let $x=$ the greater; 　　And $y=$ the less.

1. By the 1st condition, $\qquad x-y=12$
2. By the second, $\qquad x^2+y^2=1424$
3. Transp. y in the 1st, $\qquad x=y+12$
4. Squaring both sides, $\qquad x^2=y^2+24y+144$
5. Transp. y^2 in the 2d, $\qquad x^2=1424-y^2$
6. Making the 4th and 5th equal, $y^2+24y+144=1424-y^2$
7. Transp. and uniting terms, $\quad 2y^2+24y=1280$
8. Dividing by 2, $\qquad y^2+12y=640$
9. Completing the square, $\qquad y^2+12y+36=676$
10. Extracting and transp. $\qquad y=-6\pm\sqrt{676}=-6\pm26.$

$$\text{And}\quad x=y+12=20+12=32$$

EQUATIONS WHICH CONTAIN THREE OR MORE UNKNOWN QUANTITIES.

330. In the examples hitherto given, each has contained no more than *two* unknown quantities. And two independent equations have been sufficient to express the conditions of the question. But problems may involve three or more unknown quantities; and may require for their solution as many independent equations.

$$\text{Suppose } x+y+z=12$$
$$\text{And}\qquad x+2y-2z=10 \quad \Big\}\ \text{are given, to find } x, y, \text{and } z.$$
$$\text{And}\qquad x+y-z=4$$

From these three equations, two others may be derived, which shall contain only *two* unknown quantities. One of the three in the original equations may be exterminated, in the same manner as when there are, at first, only two, by the rules in arts. 324, 6, 8.

In the equations given above, if we transpose y and z, we shall have,

$$\text{In the first,}\quad x=12-y-z$$
$$\text{In the second, } x=10-2y+2z$$
$$\text{In the third,}\quad x=4-y+z$$

From these we may deduce two new equations, from which x shall be excluded.

By making the 1st and 2d equal, $\quad 12-y-z=10-2y+2z$
By making the 2d and 3d equal, $\quad 10-2y+2z=4-y+z$

By trans. and uniting terms, in the 1st of these two, $y=3z-2$
By trans. and uniting terms, in the second, $\qquad y=z+6$

From these two equations, one may be derived containing only *one* unknown quantity,

Making one equal to the other, $\quad 3z-2=z+6$
Transp. uniting terms, and dividing, $z=4$. Hence,

331. To solve a problem containing *three* unknown quantities, and producing three independent equations,

First, from the three equations deduce two, containing only two unknown quantities,

Then, from these two deduce one, containing only one unknown quantity.

For making these reductions, the rules already given are sufficient. (Art. 324, 6, 8.)

Prob. 16. Let there be given,
1. The equation $x+5y+6z=53$
2. And $\quad x+3y+3z=30$ } To find x, y, and z.
3. And $\quad x+y+z=12$

From these three equations to derive two, containing only two unknown quantities,

4. Subtract the 2d from the 1st, $\quad 2y+3z=23$
5. Subtract the 3d from the 2d, $\quad 2y+2z=18$
From these two, to derive one,
6. Subtract the 5th from the 4th, $\quad z=5$.

To find x and y, we have only to take their values from the 3d and 5th equations. (Art. 329.)

7. Transp. the 5th and dividing, $\quad y=9-z=9-5=4$
8. Transposing in the 3d, $\quad x=12-z-y=12-5-4=3$.

Prob. 17. To find x, y, and z, from
1. The equation $\quad x+y+z=12$
2. And $\quad x+2y+3z=20$
3. And $\quad \frac{1}{3}x+\frac{1}{2}y+z=6$
4. Multiplying the 1st by 3, $\quad 3x+3y+3z=36$
5. Subtracting the 2d from the 4th, $2x+y=16$
6. Subtracting the 3d from the 1st, $x-\frac{1}{3}x+y-\frac{1}{2}y=6$
7. Clearing the 6th of fractions, $\quad 4x+3y=36$
8. Multiplying the 5th by 3, $\quad 6x+3y=48$
9. Subtracting the 7th from the 8th, $2x=12$. And $x=6$.
10. Transp. in the 7th, and dividing, $y=\frac{36-4x}{3}=\frac{36-24}{3}=4$.
11. Transp. in the 1st equation, $\quad y=12-x-y=12-6-4=2$.

In this example all the reductions have been made according to the *third* rule for exterminating unknown quantities. (Art. 328.) But either of the three may be used at pleasure.

332. A calculation may often be very much abridged, by the exercise of judgment, in stating the question, in selecting the equations from which others are to be deduced, in simplifying fractional expressions, in avoiding radical quantities, &c. The skill which is necessary for this purpose, however, is to be acquired, not from a system of rules; but from practice, and a habit of attention to the peculiarities in the conditions of different problems; the variety of ways in which the same quantity may be expressed, the numerous forms which equations may assume, &c. In many of the examples in this and the preceding sections, the processes might have been shortened. But the object has been to illustrate general principles, rather than to furnish specimens of expeditious solutions. The learner will do well, as he passes along, to exercise his skill in abridging the calculations which are here given, or substituting others in their stead.

Prob. 18. Given, $\begin{cases} 1. & x+y=a \\ 2. & x+z=b \\ 3. & y+z=c \end{cases}$ To find x, y, and z.

Ans. $x=\dfrac{a+b-c}{2}$. And $y=\dfrac{a+c-b}{2}$. And $z=\dfrac{b+c-a}{2}$.

Prob. 19. Three persons A, B, and C, purchase a horse for 100 dollars, but neither is able to pay for the whole. The payment would require,

The whole of A's money, together with half of B's; or

The whole of B's, with one third of C's; or

The whole of C's, with one fourth of A's.

How much money had each?

Let $x=$ A's $z=$ C's

$y=$ B's $a=100$

1. By the first condition, $x+\frac{1}{2}y=a$ ⎞
2. By the second, $y+\frac{1}{3}z=a$ ⎬
3. By the third, $z+\frac{1}{4}x=a$ ⎠
4. Transp. in the 1st, and clear. fractions, $y=2a-2x$
5. Transp. in the second, $y=a-\frac{1}{3}z$
6. Making the 4th and 5th equal, $2a-2x=a-\frac{1}{3}z$
7. Trans. in the 6th, and clear. fractions, $z=6x-3a$ ⎞
8. Trans. in the 3d, $z=a-\frac{1}{4}x$ ⎬
9. Making the 7th and 8th equal, $6x-3a=a-\frac{1}{4}x$
10. Trans. in the 9th and clear. fractions, $25x=16a=1600$
11. Dividing by 25, $x=64$, A's money.
12 By the 8th equation, $z=a-\frac{1}{4}x=100-16=84$, C's.
13. By the 5th, $y=a-\frac{1}{3}z=100-28=72$, B's.

333. The learner must exercise his own judgment, as to the choice of the quantity to be first exterminated.   It will generally be best to begin with that which is most free from co-efficients, fractions, radical signs, &c.

Prob. 20. The sum of the distances which three persons, A, B, and C, have travelled is 62 miles;
A's distance is equal to 4 times C's, added to twice B's; and Twice A's added to 3 times B's, is equal to 17 times C's.
. What are the respective distances?
Ans. A's, 46 miles; B's, 9; and C's 7.

Prob. 21. To find x, y, and z, from
1. The equation $\frac{1}{3}x+\frac{1}{5}y+\frac{1}{2}z=62$ ⎞
2. And $\frac{1}{5}x+\frac{1}{4}y+\frac{1}{2}z=47$ ⎬
3. And $\frac{1}{4}x+\frac{1}{3}y+\frac{1}{6}z=38$ ⎠
4. Clear. the 1st of fractions, $12x+8y+6z=1488$ ⎞
5. Do.   the 2d, $20x+15y+12z=2820$ ⎬
6. Do. the 3d, $30x+24y+20z=4560$ ⎠
7. Mult. the 4th by 2, $24x+16y+12z=2976$
8. Subtract. 5th from 7th, $4x+y=156$
9. Mult. the 5th by 5, $100x+75y+60z=14100$
10. Mult. the 6th by 3, $90x+72y+60z=13680$
11. Subtract. 10th from 9th, $10x+3y=420$
12. Transp. in the 8th $y=156-4x$

13. Do. 11th, and divid. by 3, $y=\dfrac{420-10x}{3}$

14. Mak. 12th and 13th equal, $\dfrac{420-10x}{3}=156-4x$

15. Clearing of fractions, &c. $x=24$
16. By the 12th, $y=156-4x=156-96=60.$
17. By the 4th, transp. &c. $z=120.$

Prob. 22. Given $\begin{cases} xy=600 \\ xz=300 \\ yz=200 \end{cases}$ To find x, y, and z.

Ans. $x=30$. $y=20$. $z=10$.

334. The same method which is employed for the reduction of three equations, may be extended to 4, 5, or any number of equations, containing as many unknown quantities. The unknown quantities may be exterminated, one after another, and the number of equations may be reduced by successive steps, from five to four, from four to three, from three to two, &c.

Prob. 23. To find w, x, y, and z, from

1. The equation $\frac{1}{2}y+z+\frac{1}{4}w=8$ ⎫
2. And $x+y+w=9$,   ⎬ *Four* equations.
3. And $x+y+z=12$ ⎪
4. And $x+w+z=10$ ⎭
5. Clear. the 1st of frac. $y+2z+w=16$ ⎫
6. Subtract. 2d from 3d, $z-w=3$ ⎬ *Three* equations.
7. Subtract. 4th from 3d, $y-w=2$ ⎭
8. Adding 5th and 6th, $y+3z=19$ ⎫ *Two* equations.
9. Subtract. 7th from 6th, $-y+z=1$ ⎭
10. Adding 8th and 9th, $4z=20$. Or $z=5$ ⎫
11. Transp. in the 8th, $y=19-3z=19-15=4$ ⎪ Quantities
12. Transp. in the 3d, $x=12-y-z=3$ ⎬ required.
13. Transp. in the 2d, $w=9-x-y=2$ ⎭

*Prob. 24. Given $\begin{cases} w+50=x \\ x+120=3y \\ y+120=2z \\ z+195=3w \end{cases}$ To find w, x, y, and z.

Answer. $w=100$ $y=90$
                $x=150$ $z=105$.

---

335. If in the algebraic statement of the conditions of a

* For more examples of the solution of Problems by equations, see Euler's Algebra, Part. i. Sec. 4, Simpson's Algebra, Sec. ii, Simpson's Exercises, Maclaurin's Algebra, Part 1, Chap. 2 and 13, Emerson's Algebra, Book ii, Sec. 1, Saunderson's Algebra, Book ii and iii, and Dodson's Mathematical Repository.

problem, the original equations are more numerous than the unknown quantities; these equations will either be contradictory, or one or more of them will be superfluous.

Thus the equations $\begin{cases} 3x=60 \\ \frac{1}{2}x=20 \end{cases}$ are contradictory.

For by the first $x=20$, while by the second $x=40$.

. But if the latter be altered, so as to give to x the same value as the former, it will be useless, in the statement of a problem. For nothing can be determined from the one, which can not be from the other.

Thus, of the equations $\begin{cases} 3x=60 \\ \frac{1}{2}x=10 \end{cases}$ one is superfluous.

For either of them is sufficient to determine the value of x. They are not *independent* equations. (Art. 322.) One is convertible into the other. For if we divide the 1st by 6, it will become the same as the second.

Or if we multiply the second by 6, it will become the same as the first.

336. But if the number of independent equations produced from the conditions of a problem, is *less* than the number of unknown quantities, the subject is not sufficiently limited to admit of a definite answer. For each equation can limit but one quantity. And to enable us to find this quantity, all the others connected with it, must either be previously known, or be determined from other equations. If this is not the case, there will be a variety of answers which will equally satisfy the conditions of the question. If, for instance, in the equation

$$x+y=100,$$

x and y are required, there may be fifty different answers. The values of x and y may be either 99 and 1, or 98 and 2, or 97 and 3, &c. For the sum of each of these pairs of numbers is equal to 100. But if there is a second equation which determines *one* of these quantities, the other may then be found from the equation already given. As $x+y=100$, if $x=46$, y must be such a number as added to 46 will make 100, that is, it must be 54. No other number will answer this condition.

337. For the sake of abridging the solution of a problem, however, the number of independent equations actually put upon paper is frequently less, than the number of unknown quan-

·tities. Suppose we are required to divide 100 into two such parts that the greater shall be equal to three times the less. If we put x for the greater, the less will be $100-x$. (Art.195.)

Then by the supposition,	$x=300-3x$
Transposing and dividing,	$x=75$, the greater.
And	$100-75=25$, the less.

Here, two unknown quantities are found, although there appears to be but one independent equation. The reason of this is, that a part of the solution has been omitted, because it is so simple, as to be easily supplied by the mind. To have a view of the whole, without abridging, let $x=$the greater number, and $y=$the less.

1. Then by supposition,	$x+y=100$	⎫
2. And	$3y=x$	⎬
3. Transp. x in the 1st,	$y=100-x$	
4. Dividing the 2d by 3,	$y=\frac{1}{3}x$	
5. Making the 3d and 4th equal,	$\frac{1}{3}x=100-x$	
6. Multiplying by 3,	$x=300-3x$	
7. Transp. and dividing,	$x=75$, the greater.	
8. By the 3d step,	$y=100-x=25$, the less.	

By comparing these two solutions with each other, it will be seen that the first begins at the 6th step of the latter, all the preceding parts being omitted, because they are too simple to require the formality of writing down.

Prob. To find two numbers whose sum is 30, and the difference of their squares 120.

Let $a=30$ $\qquad$ $b=120$
$x=$the less number required.
Then $a-x=$the greater. (Art. 195,)
And $a^2-2ax+x^2=$the square of the greater. (Art. 214.)
From this subtract x^2 the square of the less, and we shall have $a^2-2ax=$the difference of their squares.

1. By supposition, $\qquad$ $b=a^2-2ax$
2. By transposition, $\qquad$ $2ax=a^2-b$

3. Dividing by $2a$ $\qquad$ $x=\dfrac{a^2-b}{2a}$

4. Restoring the numbers, $\qquad$ $x=\dfrac{30^2-120}{2\times30}=13$, the less.

And $a-x=30-13=17$, the greater.

X

338. In most cases also, the solution of a problem which contains many unknown quantities may be abridged, by particular artifices in *substituting* a single letter for several. (Art. 321.)

*Suppose four numbers, u, x, y, and z, are required, of which

The sum of the three first is	13
The sum of the two first and last	17
The sum of the first and two last	18
The sum of the three last	21

Then   1. $u+x+y=13$
         2. $u+x+z=17$
         3. $u+y+z=18$
         4. $x+y+z=21$

Let S be substituted for the *sum* of the four numbers, that is, for $u+x+y+z$. It will be seen that, of these four equations,

The first contains all the letters except z, that is, $S-z=13$
The second contains all except y, that is, $S-y=17$
The third contains all except x, that is, $S-x=18$
The fourth contains all except u, that is, $S-u=21$

Adding all these equations together, we have

$$4S-z-y-x-u=69$$
Or $4S-(z+y+x+u)=69$ (Art. 82.)
But $S=(z+y+x+y)$ by substitution.
Therefore, $4S-S=69$, that is, $3S=69$, and $S=23$.

Then putting 23 for S, in the four equations in which it is first introduced, we have

$$\left.\begin{array}{l} 23-z=13 \\ 23-y=17 \\ 23-x=18 \\ 23-u=21 \end{array}\right\} \text{ therefore } \left\{\begin{array}{l} z=23-13=10 \\ y=23-17=6 \\ x=23-18=5 \\ u=23-21=2 \end{array}\right.$$

Contrivances of this sort for facilitating the solution of particular problems, must be left to be furnished for the occasion, by the ingenuity of the learner. They are of a nature not to be taught by a system of rules.

339. In the resolution of equations containing several unknown quantities, there will often be an advantage in adopting the following method of notation.

*Ludlam's Algebra, art. 161. c.

•. The co-efficients of one of the unknown quantities are represented,

In the *first* equation, by a single letter, as a.

In the *second*, by the same letter marked with an accent, as a'.

In the *third*, by the same letter with a *double* accent, as a'',

&c.

The co-efficients of the other unknown quantities, are represented by other letters marked in a similar manner; as are also the terms which consist of *known* quantities only.

Two equations containing the two unknown quantities x and y may be written thus,

$$ax + by = c$$
$$a'x + b'y = c'$$

Three equations containing x, y, and z, thus,

$$ax + by + cz = d$$
$$a'x + b'y + c'z = d'$$
$$a''x + b''y + c''z = d''$$

Four equations containing x, y, z, and u, thus,

$$ax + by + cz + du = e$$
$$a'x + b'y + c'z + d'u = e'$$
$$a''x + b''y + c''z + d''u = e''$$
$$a'''x + b'''y + c'''z + d'''u = e'''$$

The same *letter* is made the co-efficient of the same unknown quantity, in different equations, that the co-efficients of the several unknown quantities may be distinguished, in any part of the calculation. But the letter is marked with different *accents*, because it actually stands for different quantities.

Thus we may put $a=4$, $a'=6$, $a''=10$, $a'''=20$, &c.

To find the value of x and y.

1. In the equation, $ax + by = c$
2. And $a'x + b'y = c'$
3. Multiplying the 1st by b',(Art.328.) $ab'x + bb'y = cb'$
4. Multiplying the 2d by b, $ba'x + bb'y = bc'$
5. Subtracting the 4th from the 3d, $ab'x - ba'x = cb' - bc'$
6. Dividing by $ab' - ba'$ (Art. 121.) $x = \dfrac{cb' - bc'}{ab' - ba'}$

By a similar process, $y = \dfrac{ac' - ca'}{ab' - ba'}$

The symmetry of these expressions is well calculated to fix them in the memory. The denominators are the same in both; and the numerators are like the denominators, except a change of one of the letters in each term. But the particular advantage of this method is, that the expressions here obtained may be considered as *general solutions*, which give the values of the unknown quantities, in other equations of a similar nature.

$$\text{Thus if } 10x+6y=100$$
$$\text{And } 40x+4y=200$$

Then putting $a=10$ $\qquad b=6$ $\qquad c=100$
$\qquad\qquad\quad a'=40$ $\qquad b'=4$ $\qquad c'=200$

We have $x=\dfrac{cb'-bc'}{ab'-ba'}=\dfrac{100\times4-6\times200}{10\times4-6\times40}=4.$

And $\qquad y=\dfrac{ac'-ca'}{ab'-ba'}=\dfrac{10\times200-100\times40}{10\times4-6\times40}=10.$

The equations to be resolved may, originally, consist of more than three terms. But if they are of the first degree, and have only two unknown quantities, each may be reduced to three terms by substitution.

Thus the equation $\qquad\qquad dx-4x+hy-6y=m+8$
Is the same, by art. 120, as $\quad (d-4)x+(h-6)y=m+8$
And putting $a=d-4,$ $\qquad b=h-6$ $\qquad c=m+8$
It becomes $\qquad\qquad\qquad ax+by=c.$*

DEMONSTRATION OF THEOREMS.

340. Equations have been applied, in this and the preceding sections, to the solution of *problems*. They may be employed with equal advantage, in the demonstration of *theorems*. The principal difference, in the two cases, is in the order in which the steps are arranged. The operations themselves are substantially the same. It is essential to a demonstration, that complete certainty be carried through every part of the pro-

---

*For the application of this plan of notation to the solution of equations which contain more than two unknown quantities, see La Croix's Algebra, art. 85, Maclaurin's Algebra, Part I. Chap. 12, Fenn's Algebra, p. 57, and a paper of Laplace, in the Memoirs of the Academy of Sciences for 1772.

cess. (Art.11.) This is effected, in the reduction of equations, by adhering to the general rule, to make no alteration which shall affect the value of one of the members, without equally increasing or diminishing the other. In applying this principle, we are guided by the axioms laid down in art. 63. These axioms are as applicable to the demonstration of theorems, as to the solution of problems.

But the *order* of the steps will generally be different. In solving a problem, the object is to find the value of the unknown quantity, by disengaging it from all other quantities. But in conducting a demonstration, it is necessary to bring the equation to that particular form which will express, in algebraic terms, the proposition to be proved.

Ex. 1. *Theorem.* Four times the product of any two numbers, is equal to the square of their sum, diminished by the square of their difference.

Let $x=$ the greater number, $s=$ their sum,
     $y=$ the less, $d=$ their difference?

Demonstration.

1. By the notation $x+y=s$ }
2. And $x-y=d$ }
3. Adding the two, (Ax. 1.) $2x=s+d$
4. Subtracting the 2d from the 1st, $2y=s-d$
5. Mult. 3d and 4th, (Ax. 3.) $4xy=(s+d)\times(s-d)$
6. That is, (Art. 235.) $4xy=s^2-d^2.$

The last equation expressed in words is the proposition which was to be demonstrated. It will be easily seen that it is equally applicable to any two numbers whatever. For the particular values of x and y will make no difference in the nature of the proof.

Thus $4\times8\times6=(8+6)^2-(8-6)^2=192.$
And $4\times10\times6=(10+6)^2-(10-6)^2=240.$
And $4\times12\times10=(12+10)^2-(12-10)^2=480.$

Theorem 2. The sum of the squares of any two numbers, is equal to the square of their difference, added to twice their product.

Let $x=$ the greater, $d=$ their difference.
     $y=$ the less, $p=$ their product.

Demonstration.

1. By the notation $\quad\quad\quad x-y=d \left.\right\}$
2. And $\quad\quad\quad\quad\quad\quad xy=p$
3. Squaring the first, $\quad\quad x^2-2xy+y^2=d^2$
4. Multiplying the 2d by 2, $\quad 2xy=2p$
5. Adding the 3d and 4th, $\quad x^2+y^2=d^2+2p$.

Thus $10^2+8^2=(10-8)^2+2\times10\times8=164$.

341. General propositions are also *discovered*, in an expeditious manner, by means of equations. The relations of quantities may be presented to our view, in a great variety of ways, by the several changes through which a given equation may be made to pass. Each step in the process will contain a distinct proposition.

Let s and d be the sum and difference of two quantities x and y, as before.

1. Then $\quad\quad\quad\quad\quad s=x+y \left.\right\}$
2. And $\quad\quad\quad\quad\quad\quad d=x-y$
3. Dividing the 1st by 2, $\quad \frac{1}{2}s=\frac{1}{2}x+\frac{1}{2}y$
4. Dividing the 2d by 2, $\quad \frac{1}{2}d=\frac{1}{2}x-\frac{1}{2}y$
5. Adding the 3d and 4th, $\quad \frac{1}{2}s+\frac{1}{2}d=\frac{1}{2}x+\frac{1}{2}x=x$
6. Sub. the 4th from the 3d, $\quad \frac{1}{2}s-\frac{1}{2}d=\frac{1}{2}y+\frac{1}{2}y=y$.

That is,

Half the difference of two quantities, added to half their sum, is equal to the greater; and

Half their difference subtracted from half their sum, is equal to the less.

RATIO AND PROPORTION.*

ART. 342. THE design of mathematical investigations, is to arrive at the knowledge of particular quantities, by comparing them with other quantities, either *equal to*, or, *greater*, or *less* than those which are the objects of inquiry. The end is most commonly attained by means of a series of *equations* and *proportions*. When we make use of equations, we determine the quantity sought, by discovering its *equality* with some other quantity or quantities already known.

We have frequent occasion, however, to compare the unknown quantity with others which are *not equal* to it, but either greater or less. Here, a different mode of proceeding becomes necessary. We may inquire, either *how much* one of the quantities is greater than the other; or *how many times* the one contains the other. In finding the answer to either of these inquiries, we discover what is termed a *ratio* of the two quantities. One is called *arithmetical*, and the other *geometrical* ratio. It should be observed, however, that both these terms have been adopted arbitrarily, merely for distinction sake. Arithmetical ratio, and geometrical ratio, are both of them applicable to arithmetic, and both to geometry.

As the whole of the extensive and important subject of proportion depends upon ratios, it is necessary that these should be clearly and fully understood.

343. ARITHMETICAL RATIO *is the* DIFFERENCE *between two quantities or sets of quantities.* The quantities themselves are called the *terms* of the ratio, that is, the terms between

---

* Euclid's Elements, Book 5, 7, 8. Euler's Algebra, Part I. Sec. 3. Emerson on Proportion. Camus' Geometry, Book III. Ludlam's Mathematics. Wallis' Algebra, Chap. 19, 20. Saunderson's Algebra, Book 7. Barrow's Mathematical Lectures. See also an ingenious essay on the 5th book of Euclid, in the Analyst for March 1814, by Professor Adrain.

which the ratio exists. Thus 2 is the arithmetical ratio of 5 to 3. This is sometimes expressed, by placing two points between the quantities thus $5 . . 3$, which is the same as $5-3$. Indeed the term arithmetical ratio, and its notation by points are almost needless. For the one is only a substitute for the word *difference*, and the other for the sign $-$.

344. If both the terms of an arithmetical ratio be *multiplied* or *divided* by the same quantity, the *ratio* will, in effect, be multiplied or divided by that quantity.

Thus if $\qquad a-b=r$

Then mult. both sides by h, (Ax. 3.) $\quad ha-hb=hr$

And dividing by h, (Ax. 4.) $\qquad \dfrac{a}{h}-\dfrac{b}{h}=\dfrac{r}{h}$.

345. If the terms of one arithmetical ratio be added to, or subtracted from, the corresponding terms of another, the ratio of their sum or difference will be equal to the sum or difference of the two ratios.

$$\left.\begin{array}{l} \text{If} \quad a-b \\ \text{And} \quad d-h \end{array}\right\} \text{are the two ratios,}$$

Then $(a+d)-(b+h)=(a-b)+(d-h)$. For each $=a+d-b-h$.
And $(a-d)-(b-h)=(a-b)-(d-h)$. For each $=a-d-b+h$.

$$\left.\begin{array}{l} \text{Thus the arith. ratio of } 11..4 \text{ is } 7 \\ \text{And the arith. ratio of } 5..2 \text{ is } 3 \end{array}\right\}$$

The ratio of the sum of the terms $16..6$ is 10, the sum of the ratios. The ratio of the diff. of the terms $6..2$ is 4, the diff. of the ratios.

346. GEOMETRICAL RATIO *is that relation between quantities which is expressed by the* QUOTIENT *of the one divided by the other.*

Thus the ratio of 8 to 4, is $\frac{8}{4}$ or 2. For this is the quotient of 8 divided by 4. In other words, it shows how often 4 is contained in 8.

In the same manner, the ratio of any quantity to another may be expressed by dividing the former by the latter, or, which is the same thing, making the former the numerator of a fraction, and the latter the denominator.

Thus the ratio of a to b is $\dfrac{a}{b}$.

The ratio of $d+h$ to $b+c$, is $\dfrac{d+h}{b+c}$.

347. Geometrical ratio is also expressed by placing two points, one over the other, between the quantities compared.

Thus $a : b$ expresses the ratio of a to b; and $12 : 4$ the ratio of 12 to 4. The two quantities together are called a *couplet*, of which the first term is the *antecedent*, and the last, the *consequent*.

348. This notation by points, and the other in the form of a fraction, may be exchanged the one for the other, as convenience may require; observing to make the antecedent of the couplet, the numerator of the fraction, and the consequent the denominator.

Thus $10 : 5$ is the same as $\frac{10}{5}$ and $b : d$, the same as $\frac{b}{d}$.

349. Of these three, the antecedent, the consequent, and the ratio, any *two* being given, the other may be found.

Let $a =$ the antecedent, $c =$ the consequent, $r =$ the ratio.

By definition $r = \frac{a}{c}$; that is, the ratio is equal to the antecedent divided by the consequent.

Multiplying by c, $a = cr$, that is, the antecedent is equal to the consequent multiplied into the ratio.

Dividing by r, $c = \frac{a}{r}$, that is, the consequent is equal to the antecedent divided by the ratio.

Cor. 1. If two couplets have their antecedents equal, and their consequents equal, their ratios must be equal.*

Cor. 2. If, in two couplets, the ratios are equal, and the antecedents equal, the consequents are equal; and if the ratios are equal and the consequents equal, the antecedents are equal.†

350. If the two quantities compared are *equal*, the ratio is a unit, or a ratio of equality. Thus the ratio of $3 \times 6 : 18$ is a unit, for the quotient of any quantity divided by itself is 1.

If the antecedent of a couplet is *greater* than the consequent, the ratio is greater than a unit. For if a dividend is greater than its divisor, the quotient is greater than a unit.

* Euclid 7. 5.   † Euc. 9, 5.

Y

Thus the ratio of 18 : 6 is 3. (Art. 128. cor.) This is called a ratio of *greater inequality*.

On the other hand, if the antecedent is *less* than the consequent, the ratio is less than a unit, and is called a ratio of *less inequality*. Thus the ratio of 2 : 3, is less than a unit; because the dividend is less than the divisor.

351. INVERSE *or* RECIPROCAL *ratio is the ratio of the reciprocals of two quantities.* See art. 49.

Thus the reciprocal ratio of 6 to 3, is $\frac{1}{6}$ to $\frac{1}{3}$, that is $\frac{1}{6} \div \frac{1}{3}$.

The direct ratio of a to b is $\frac{a}{b}$, that is, the antecedent divided by the consequent.

The reciprocal ratio, is $\frac{1}{a} : \frac{1}{b}$ or $\frac{1}{a} \div \frac{1}{b} = \frac{1}{a} \times \frac{b}{1} = \frac{b}{a}$, that is, the consequent b divided by the antecedent a.

Hence a reciprocal ratio is expressed by *inverting the fraction* which expresses the direct ratio; or, when the notation is by points, by *inverting the order of the terms.*

Thus a is to b, *inversely,* as b to a.

352. COMPOUND RATIO *is the ratio of the* PRODUCTS *of the corresponding terms of two or more simple ratios.**

Thus the ratio of   6 : 3, is 2
And the ratio of   12 : 4, is 3
_____
The ratio compounded of these is   72 : 12 = 6.

Here the compound ratio is obtained by multiplying together the two antecedents, and also the two consequents, of the simple ratios.

So the ratio compounded,

Of the ratio of $a : b$
And the ratio of $c : d$
And the ratio of $h : y$
_____

Is the ratio of $ach : bdy = \dfrac{ach}{bdy}$.

Compound ratio is not different in its *nature* from any other ratio. The term is used, to denote the origin of the ratio, in particular cases.

*See note D.

Cor. The compound ratio is equal to the product of the simple ratios.

The ratio of $a:b,$ is $\dfrac{a}{b}$

The ratio of $c:d,$ is $\dfrac{c}{d}$

The ratio of $h:y,$ is $\dfrac{h}{y}$

And the ratio compound of these is $\dfrac{ach}{bdy}$, which is the product of the fractions expressing the simple ratios. (Art. 155.)

So the ratio of $\quad$ $8:4$ is 2
The ratio of $\quad$ $6:2$ is 3
The ratio of $\quad$ $8:2$ is 4

And the ratio compounded of these is $24 = 2 \times 3 \times 4$.

353. If, in a series of ratios, the consequent of each preceding couplet, is the antecedent of the following one, *the ratio of the first antecedent to the last consequent, is equal to that which is compounded of all the intervening ratios.*[*]

Thus, in the series of ratios $\quad a:b$
$b:c$
$c:d$
$d:h$

the ratio of $a:h$ is equal to that which is compounded of the ratios of $a:b$, of $b:c$, of $c:d$, of $d:h$. For the compound ratio, by the last article, is $\dfrac{abcd}{bcdh} = \dfrac{a}{h}$, or $a:h$. (Art. 145.)

In the same manner, all the quantities which are both antecedents and consequents will *disappear* when the fractional product is reduced to its lowest terms, and will leave the compound ratio to be expressed by the first antecedent and the last consequent.

The ratio compounded of $2:6$
$6:8$
$8:15$, is $\dfrac{96}{720} = \dfrac{2}{15}$ or $2:15$.

---

[*] This is the particular case of compound ratio which is treated of in the 5th book of Euclid. See the editions of Simson and Playfair.

354. A particular class of compound ratios is produced, by multiplying a simple ratio into *itself,* or into another *equal* ratio. These are termed *duplicate, triplicate, quadruplicate,* &c. according to the number of multiplications.

A ratio compounded of *two* equal ratios, that is, the *square* of the simple ratio, is called a *duplicate* ratio.

One compounded of *three,* that is, the *cube* of the simple ratio, is called *triplicate,* &c.

Thus the simple ratio of a to b, is $a : b = \dfrac{a}{b}$

The duplicate ratio of a to b, is $a^2 : b^2 = \dfrac{a^2}{b^2}$

The triplicate ratio of a to b, is $a^3 : b^3 = \dfrac{a^3}{b^3}$, &c.

The terms *duplicate, triplicate,* &c. ought not to be confounded with *double, triple,* &c.*

The ratio of 6 to 2 is $6 : 2 = 3$

Double this ratio, that is, *twice* the ratio is $12 : 2 = 6$ }

Triple the ratio, i. e. *three times* the ratio, is $18 : 2 = 9$ }

But the *duplicate* ratio, i. e. the *square* of the ratio, is $6^2 : 2^2 = 9$ }

And the *triplicate* ratio, i. e. the *cube* of the ratio, is $6^3 : 2^3 = 27$ }

355. That quantities may have a ratio to each other, it is necessary that they should be so far of the same nature, as that one can properly be said to be either equal to, or greater, or less than the other. A foot has a ratio to an inch, for one is twelve times as great as the other. But it can not be said that an hour is either longer or shorter than a rod; or' that an acre is greater or less than a degree. Still, if these quantities are expressed by *numbers,* there may be a ratio between the numbers. There is a ratio between the number of minutes in an hour, and the number of rods in a mile.

356. Having attended to the *nature* of ratios, we have next to consider in what manner they will be affected, by varying one or both of the terms between which the comparison is made. It must be kept in mind that, when a direct ratio is expressed by a fraction, the *antecedent* of the couplet is always the *numerator,* and the *consequent,* the *denominator.* It will be easy, then, to derive from the properties of fractions, the changes produced in ratios by variations in the quantities compared. For the ratio of the two quantities is the same as the *value* of the fractions, each be-

* See Note E.

ing the *quotient* of the numerator divided by the denominator. (Arts. 135, 346.) Now it has been shown, (Art. 137.) that multiplying the numerator of a fraction by any quantity is multiplying the *value* by that quantity; and that dividing the numerator is dividing the value. Hence,

357. *Multiplying the antecedent of a couplet by any quantity, is multiplying the ratio by that quantity ; and dividing the antecedent is dividing the ratio.*

Thus the ratio of   6 : 2 is 3
And the ratio of   24 : 2 is 12.

Here the antecedent and the ratio, in the last couplet, are each four times as great as in the first.

The ratio of $a : b$ is $\dfrac{a}{b}$

And the ratio of $na : b$ is $\dfrac{na}{b}$.

Cor. With a given consequent, the greater the *antecedent*, the greater the *ratio;* and on the other hand, the greater the ratio, the greater the antecedent.* See art. 137. cor.

358. *Multiplying the consequent of a couplet by any quantity is, in effect, dividing the ratio by that quantity; and dividing the consequent is multiplying the ratio.* For multiplying the denominator of a fraction, is dividing the value ; and dividing the denominator is multiplying the value. (Art.138.)

Thus the ratio of 12 : 2, is 6
And the ratio of 12 : 4, is 3.

Here the consequent, in the second couplet, is *twice* as great, and the ratio only *half* as great, as in the first.

The ratio of $a : b$, is $\dfrac{a}{b}$

And the ratio of $a : nb$, is $\dfrac{a}{nb}$.

Cor. With a given antecedent, the greater the consequent, the less the ratio; and the greater the ratio, the less the consequent.† See art. 138. cor.

359. From the two last articles, it is evident that *multiplying the antecedent* of a couplet, by any quantity, will have the

* Euclid 8 and 10. 5. The first part of the propositions.
† Euclid 8 and 10. 5. The last part of the propositions.

same effect on the ratio, as *dividing the consequent*, by that quantity; and *dividing the antecedent* will have the same effect as *multiplying the consequent*. See art. 139.

Thus the ratio of 8:4, is 2
Mult. the antecedent by 2, the ratio of 16:4, is 4
Divid. the consequent by 2, the ratio of 8:2, is 4.

Cor. Any *factor* or *divisor* may be transferred, from the antecedent of a couplet to the consequent, or from the consequent to the antecedent, without altering the ratio.

It must be observed that, when a factor is thus transferred from one term to the other, it becomes a divisor; and when a divisor is transferred, it becomes a factor.

Thus the ratio of $3 \times 6 : 9 = 2$
Transferring the factor 3, $6 : \frac{9}{3} = 2$ the same ratio.

The ratio of

$$\frac{ma}{y} : b = \frac{ma}{y} \div b = \frac{ma}{by}$$

Transferring y,

$$ma : by = ma \div by = \frac{ma}{by}$$

Transferring m,

$$a : \frac{by}{m} = a \div \frac{by}{m} = \frac{ma}{by}$$

360. It is farther evident, from arts. 357 and 358, that *if the antecedent and consequent be* BOTH *multiplied, or both divided, by the same quantity, the ratio will not be altered.*[*] See art. 140.

Thus the ratio of $8:4=2$
Mult. both terms by 2, $16:8=2$ the same ratio.
Divid. both terms by 2, $4:2=2$

The ratio of

$$a:b = \frac{a}{b}$$

Mult. both terms by m,

$$ma:mb = \frac{ma}{mb} = \frac{a}{b}$$

Divid. both terms by n,

$$\frac{a}{n} \cdot \frac{b}{n} = \frac{an}{bn} = \frac{a}{b}$$

Cor. 1. The *halves* of quantities have the same ratio as their wholes.

[*] Euclid 15. 5.

Cor. 2. The ratio of two *fractions* which have a common denominator, is the same as the ratio of their *numerators*.

The ratio of $\frac{a}{n} : \frac{b}{n}$, is the same as that of $a : b$.

Cor. 3. The *direct* ratio of two fractions which have a common numerator, is the same as the reciprocal ratio of their *denominators*.

Thus the ratio of $\frac{a}{m} : \frac{a}{n}$, is the same as $\frac{1}{m}$ $\frac{1}{n}$ or $n : m$.

361. From the last article, it will be easy to determine the ratio of any two fractions. If each term be multiplied by the two denominators, the ratio will be assigned in integral expressions. Thus, multiplying the terms of the couplet $\frac{a}{b} : \frac{c}{d}$ by bd, we have $\frac{abd}{b} : \frac{bcd}{d}$, which becomes $ad : bc$, by cancelling equal quantities from the numerators and denominators.

362. *If to or from the terms of any couplet, there be* ADDED *or* SUBTRACTED *two other quantities having the same ratio, the sums or remainders will also have the same ratio.**

> Let the ratio of $\qquad a : b$ ⎫
> Be the same as that of $\qquad c : d$ ⎬

Then the ratio of the *sum* of the antecedents, to the sum of the consequents, viz. of $a+c$ to $b+d$, is also the same.

That is $\dfrac{a+c}{b+d} = \dfrac{c}{d} = \dfrac{a}{b}$.

Demonstration.

1. By supposition, $\qquad\qquad \dfrac{a}{b} = \dfrac{c}{d}$

2. Mult. by b and d, (Ax. 3.) $\qquad ad = bc$

3. Adding cd to both sides, (Ax. 1.) $ad+cd = bc+cd$

4. Dividing by d, (Ax. 4.) $\qquad a+c = \dfrac{bc+cd}{d}$

5. Dividing by $b+d$, (Art. 121.) $\dfrac{a+c}{b+d} = \dfrac{c}{d} = \dfrac{a}{b}$

* Euclid 5 and 6. 5.

The ratio of the *difference* of the antecedents, to the difference of the consequents, is also the same.

That is $\dfrac{a-c}{b-d}=\dfrac{c}{d}=\dfrac{a}{b}$.

Demonstration.

1. By supposition, as before, $\dfrac{a}{b}=\dfrac{c}{d}$

2. Multiplying by b and d, $ad=bc$

3. Subtracting cd from both sides, $ad-cd=bc-cd$

4. Dividing by d, $a-c=\dfrac{bc-cd}{d}$

5. Dividing by $b-d$ $\dfrac{a-c}{b-d}=\dfrac{c}{d}=\dfrac{a}{b}$.

Thus the ratio of   15:5 is 3 }
And the ratio of   9:3 is 3 }

Then adding and subtracting the terms of the two couplets.

The ratio of   15+9:5+3 is 3 }
And the ratio of   15−9:5−3 is 3 }

Here the terms of only *two* couplets have been added together. But the proof may be extended to *any number* of couplets, where the ratios are equal. For, by the addition of the two first, a *new* couplet is formed, to which, upon the same principle, a third may be added, a fourth, &c. Hence,

363. If, in several couplets, the ratios are equal, *the sum of all the antecedents has the same ratio to the sum of all the consequents, which any one of the antecedents has, to its consequent.**

Thus the ratio of $\begin{cases} 12:6=2 \\ 10:5=2 \\ 8:4=2 \\ 6:3=2 \end{cases}$

Therefore the ratio of $(12+10+8+6):(6+5+4+3)=2$.

* Euclid 1 and 12. 5.

PROPORTION.

363. An accurate and familiar acquaintance with the doctrine of ratios, is necessary to a ready understanding of the principles of *proportion*, one of the most important of all the branches of the mathematics. In considering ratios, we compare two *quantities*, for the purpose of finding either their difference, or the quotient of the one divided by the other. But in proportion, the comparison is between two *ratios*. And this comparison is limited to such ratios as are *equal*. We do not inquire how much one ratio is *greater* or *less* than another, but whether they are the *same*. Thus the numbers 12, 6, 8, 4, are said to be proportional, because the ratio of 12:6 is the same as that of 8:4.

364. PROPORTION, then, *is an equality of ratios*. It is either *arithmetical* or *geometrical*. Arithmetical proportion is an equality of arithmetical ratios, and geometrical proportion is an equality of geometrical ratios.* Thus the numbers 6, 4, 10, 8, are in *arithmetical* proportion, because the *difference* between 6 and 4 is the same as the difference between 10 and 8. And the numbers 6, 2, 12, 4, are in *geometrical* proportion, because the *quotient* of 6 divided by 2 is the same, as the quotient of 12 divided by 4.

365. Care must be taken not to confound *proportion* with *ratio*. This caution is the more necessary, as in common discourse, the two terms are used indiscriminately, or rather, proportion is used for both. The expenses of one man are said to bear a greater proportion to his income, than those of another. But according to the definition which has just been given, one proportion is neither greater nor less than another. For *equality* does not admit of degrees. One *ratio* may be greater or less than another. The ratio of 12:2 is greater than that of 6:2, and less than that of 20:2. But these differences are not applicable to *proportion*, when the term is used in its technical sense. The loose signification which is so frequently attached to this word may be proper enough in *familiar language*. For it is sanctioned by general usage. But, for scientific purposes, the distinction between proportion and ratio, should be clearly drawn, and cautiously observed.

* See Note F.

Z

366. The equality between two ratios, as has been stated, is called proportion. The word is sometimes applied also to the series of terms among which this equality of ratios exists. Thus the two couplets $15:5$ and $6:2$ are, when taken together, called a proportion.

367. Proportion may be expressed, either by the common sign of equality, or by four points between the two couplets.

Thus $\begin{cases} 8 \cdot\cdot 6 = 4 \cdot\cdot 2, \text{ or } 8 \cdot\cdot 6 :: 4 \cdot\cdot 2 \\ a \cdot\cdot b = c \cdot\cdot d, \text{ or } a \cdot\cdot b :: c \cdot\cdot d \end{cases}$ are arithmetical proportions.

And $\begin{cases} 12:6 = 8:4, \text{ or } 12:6::8:4 \\ a:b = d:h, \text{ or } a:b::d:h \end{cases}$ are geometrical proportions.

The latter is read, 'the ratio of a to b equals the ratio of d to h;' or more concisely, 'a is to b, as d to h.'

368. The first and last terms are called the *extremes*, and the other two the *means*. *Homologous* terms are either the two antecedents or the two consequents. *Analogous* terms are the antecedent and consequent of the same couplet.

369. As the ratios are equal, it is manifestly immaterial which of the two couplets is placed first.

If $a:b::c:d$, then $c:d::a:b$. For if $\dfrac{a}{b}=\dfrac{c}{d}$ then $\dfrac{c}{d}=\dfrac{a}{b}$.

370. The number of terms must be, at least, four. For the equality is between the ratios of *two couplets;* and each couplet must have an antecedent and a consequent. There may be a proportion, however, among three *quantities.* For one of the quantities may be *repeated,* so as to form two terms. In this case, the quantity repeated is called the *middle term*, or a *mean proportional* between the two other quantities, especially if the proportion is geometrical.

Thus the numbers 8, 4, 2, are proportional. That is, $8:4::4:2$. Here 4 is both the consequent in the first couplet, and the antecedent in the last. It is therefore a mean proportional between 8 and 2.

The *last* term is called a *third proportional* to the two other quantities. Thus 2 is a third proportional to 8 and 4.

371. *Inverse* or *reciprocal* proportion is an equality between a *direct* ratio and a *reciprocal* ratio.

Thus $4:2::\frac{1}{3}:\frac{1}{6}$; that is, 4 is to 2, *reciprocally*, as 3 to 6. Sometimes also, the order of the terms in one of the couplets is inverted, without writing them in the form of a fraction. (Art. 351.)

Thus $4:2::3:6$ inversely. In this case, the *first* term is to the *second*, as the *fourth* to the *third;* that is, the first divided by the second, is equal to the fourth divided by the third.

372. When there is a series of quantities, such that the ratios of the first to the second, of the second to the third, of the third to the fourth, &c. are *all equal ;* the quantities are said to be in *continued proportion.* The consequent of each preceding ratio is, then, the antecedent of the following one. Continued proportion is also called *progression,* as will be seen in a following section.

Thus the numbers 10, 8, 6, 4, 2, are in continued *arithmetical* proportion. For $10-8=8-6=6-4=4-2$.

The numbers 64, 32, 16, 8, 4, are in continued *geometrical* proportion. For $64:32::32:16::16:8::8:4$.

If a, b, c, d, h, &c. are in continued geometrical proportion; then $a:b::b:c::c:d::d:h$, &c.

One case of continued proportion is that of *three* proportional quantitities. (Art. 370.)

373. As *arithmetical* proportion is, generally, nothing more than a very simple equation, it is scarcely necessary to treat of it separately.

The proportion $\qquad a \cdot\cdot b :: c \cdot\cdot d$
Is the same as the equation $\qquad a-b=c-d.$

It will be proper, however, to observe that, if *four* quantities are in arithmetical proportion, *the sum of the extremes is equal to the sum of the means.*

Thus if $\qquad a \cdot\cdot b :: h \cdot\cdot m$, then $\qquad a+m=b+h$
For by supposition, $\qquad a-b=h-m$
And transp.$-b$ and $-m$ $\qquad a+m=b+h.$

So in the proportion,$12 \cdot\cdot 10 :: 11 \cdot\cdot 9$,we have $12+9=10+11.$

Again, if *three* quantities are in arithmetical proportion, *the sum of the extremes is equal to double the mean.*

If $a \cdot\cdot b :: b \cdot\cdot c$, then $\qquad a-b=b-c$
And transposing $-b$ and $-c$, $\qquad a+c=2b.$

GEOMETRICAL PROPORTION.

374. But if four quantities are in *geometrical* proportion, *the* PRODUCT *of the extremes is equal to the product of the means.*

If $a:b::c:d$, then $ad=bc$.

For by supposition, (Arts. 346, 364.) $\dfrac{a}{b}=\dfrac{c}{d}$.

Multiplying by bd, (Ax. 3.) $\dfrac{abd}{b}=\dfrac{cbd}{d}$

Reducing the fractions, $ad=bc.$

Thus $12:8::15:10$, therefore $12\times10=8\times15.$

375. On the other hand, if the product of two quantities is equal to the product of two others, the four quantities will form a proportion, when they are so arranged, that those on one side of the equation shall constitute the means, and those on the other side, the extremes.

If $my=nh$, then $m:n::h:y$, that is $\dfrac{m}{n}=\dfrac{h}{y}.$

For by dividing $my=nh$ by ny, we have $\dfrac{my}{ny}=\dfrac{nh}{ny}$

And reducing the fractions, $\dfrac{m}{n}=\dfrac{h}{y}.$

Cor. The same must be true of any factors which form the two sides of an equation.

If $(a+b)\times c=(d-m)\times y$, then $a+b:d-m::y:c.$

376. If three quantities are proportional, the product of the extremes is equal to the square of the means. For this mean proportional is, at the same time, the consequent of the first couplet, and the antecedent of the last. (Art. 370.) It is therefore to be multiplied into itself, that is, it is to be squared.

If $a:b::b:c$, then mult. extremes and means, $ac=b^2.$

377. It follows from art. 374, that in a proportion, either extreme is equal to the product of the means, divided by the other extreme; and either of the means is equal to the product of the extremes, divided by the other mean.

1. If $a:b::c:d$, then $\qquad$ $ad=bc$

2. Dividing by d, $\qquad$ $a=\dfrac{bc}{d}$

3. Dividing the first by c, $\quad$ $b=\dfrac{ad}{c}$

4. Dividing it by b, $\qquad$ $c=\dfrac{ad}{b}$

5. Dividing it by a, $\qquad$ $d=\dfrac{bc}{a}$; that is, the *fourth* term is equal to the *product of the second and third divided by the first.*

On this principle is founded the rule of simple proportion in arithmetic, commonly called the *Rule of Three.* Three numbers are given to find a fourth, which is obtained by multiplying together the second and third, and dividing by the first.

378. The propositions respecting the products of the means, and of the extremes, furnish a very simple and convenient criterion for determining whether any four quantities are proportional. We have only to multiply the means together, and also the extremes. If the two products are equal, the quantities are proportional. If the products are not equal, the quantities are not proportional.

379. In mathematical investigations, when the relations of several quantities are given, they are frequently stated in the form of a proportion. But it is commonly necessary that this first proportion should pass through a number of transformations, before it brings out distinctly the unknown quantity, or the proposition which we wish to demonstrate. It may undergo any change which will not affect the equality of the ratios; or which will leave the product of the means equal to the product of the extremes.

It is evident, in the first place, that any alteration in the *arrangement*, which will not affect the equality of these two products, will not destroy the proportion. Thus, if $a:b::c:d$, the order of these four quantities may be varied, in any way which will leave $ad=bc$. Hence,

380. If four quantities are proportional, *the order of the means, or of the extremes, or of the terms of both couplets, may be inverted, without destroying the proportion.*

$$\text{If } \quad a : b :: c : d \atop \text{And } 12 : 8 :: 6 : 4 \Big\} \text{ then,}$$

1. *Inverting the means,*[*]

$$a : c :: b : d \atop 12 : 6 :: 8 : 4 \Big\} \text{ that is, } \Big\{ \begin{array}{l} \text{The } \textit{first, } \text{is to the } \textit{third,} \\ \text{As the } \textit{second, } \text{to the } \textit{fourth.} \end{array}$$

In other words, the ratio of the *antecedents* is equal to the ratio of the *consequents*.

This inversion of the means is frequently referred to by geometers under the name of *Alternation.*[†]

2. *Inverting the extremes.*

$$d : b :: c : a \atop 4 : 8 :: 6 : 12 \Big\} \text{ that is, } \Big\{ \begin{array}{l} \text{The } \textit{fourth, } \text{is to the } \textit{second,} \\ \text{As the } \textit{third, } \text{to the } \textit{first.} \end{array}$$

3. *Inverting the terms of each couplet.*

$$b : a :: d : c \atop 8 : 12 :: 4 : 6 \Big\} \text{ that is, } \Big\{ \begin{array}{l} \text{The } \textit{second, } \text{is to the } \textit{first,} \\ \text{As the } \textit{fourth, } \text{to the } \textit{third.} \end{array}$$

This is technically called *Inversion.*

Each of these may also be varied, by changing the order of the *two couplets.* (Art. 369.)

Cor. The order of the *whole proportion* may be inverted. If $a : b :: c : d$, then $d : c :: b : a$.

In each of these cases, it will be at once seen that, by taking the products of the means, and of the extremes, we have $ad = bc$, and $12 \times 4 = 8 \times 6$.

If the terms of only *one* of the couplets are inverted, the proportion becomes *reciprocal.* (Art. 371.)

If $a : b :: c : d$, then a is to b, reciprocally as d to c.

381. A difference of arrangement is not the *only* alteration which we have occasion to produce, in the terms of a proportion. It is frequently necessary to multiply, divide, involve, &c. In all cases, the art of conducting the investigation consists in so ordering the several changes, as to maintain a constant equality, between the ratio of the two first terms, and that of the two last. As in resolving an equation, we must see that the *sides* remain equal; so in varying a proportion, the equality of the *ratios* must be preserved. And this is effected, either by keeping the ratios the *same,*

---

[*] See Note G.   [†] Euclid 16. 5.

while the *terms* are altered; or by increasing or diminishing one of the ratios, *as much as the other.* Most of the succeeding proofs are intended to bring this principle distinctly into view, and to make it familiar. Some of the propositions might be demonstrated, in a more simple manner, perhaps, by multiplying the extremes and means. But this would not give so clear a view of the *nature* of the several changes in the proportions.

It has been shown that, if *both* the terms of a couplet be multiplied or divided by the same quantity, the ratio will remain the same; (Art. 360.) that multiplying the *antecedent* is, in effect, multiplying the ratio, and dividing the antecedent is dividing the ratio; (Art. 357.) and farther, that multiplying the *consequent* is, in effect, dividing the ratio, and dividing the consequent is multiplying the ratio. (Art. 358.) As the ratios in a proportion are equal, if they are both multiplied, or both divided, by the same quantity, they will still be equal. (Ax. 3.) One will be increased or diminished as much as the other. Hence,

382. If four quantities are proportional, *two analogous, or two homologous terms may be multiplied or divided by the same quantity, without destroying the proportion.*

If *analogous* terms be multiplied or divided, the ratios will not be altered. (Art. 360.) If *homologous* terms be multiplied or divided, both ratios will be equally increased or diminished. (Arts. 357, 8.)

$$\text{If} \quad a:b::c:d \atop \text{And } 12:8::6:4 \Big\} \text{ then,}$$

1. *Multiplying the two first terms,*

$$ma: mb :: c:d \atop 2\times12:2\times8::6:4 \Big\} \text{ The ratios not altered.}$$

Multiplying the two last terms.

$$a:b:: mc :md \atop 12:8::2\times6:2\times4 \Big\} \text{ The ratios not altered.}$$

3. *Multiplying the two antecedents.**

$$ma:b:: mc :d \atop 2\times12:8::2\times6:4 \Big\} \text{ Both ratios equally increased.}$$

* Euclid 3. 5.

4. *Multiplying the two consequents.*

$$\left.\begin{array}{l} a:mb::c:md \\ 12:2\times8::6:2\times4 \end{array}\right\} \text{Both ratios equally diminished.}$$

5. *Dividing the two first terms.*

$$\frac{a}{m}:\frac{b}{m}::c:d. \quad \text{The ratios not altered.}$$

6. *Dividing the two last terms.*

$$a:b::\frac{c}{m}:\frac{d}{m}. \quad \text{The ratios not altered.}$$

7. *Dividing the two antecedents.*

$$\frac{a}{m}:b::\frac{c}{m}:d. \quad \text{The ratios equally diminished.}$$

8. *Dividing the two consequents.*

$$a:\frac{b}{m}::c:\frac{d}{m}. \quad \text{The ratios equally increased.}$$

Cor. 1. *All* the terms may be multiplied or divided by the same quantity.*

$$\left.\begin{array}{l} ma:mb::mc:md \\ \dfrac{a}{m}:\dfrac{b}{m}::\dfrac{c}{m}:\dfrac{d}{m} \end{array}\right\} \text{The ratios not altered.}$$

Cor. 2. If four quantities are proportional, their *halves* are proportional.

Cor. 3. In any of the cases in this article, multiplication of the consequent may be substituted for division of the antecedent in the same couplet, and division of the consequent, for multiplication of the antecedent. (Art. 359, cor.)

$$\text{Thus for} \left\{\begin{array}{l} ma:b::mc:d \\ \dfrac{a}{m}:b::\dfrac{c}{m}:d \end{array}\right\} \text{may be put} \left\{\begin{array}{l} a:\dfrac{b}{m}::mc:d \\ a:mb::\dfrac{c}{m}:d \end{array}\right\} \text{or} \left\{\begin{array}{l} ma:b::c:\dfrac{d}{m} \\ \dfrac{a}{m}:b::c:md. \end{array}\right.$$

* Euclid 4. 5.

383. It is often necessary, not only to alter the terms of a proportion, and to vary the arrangement, but to *compare one proportion with another.* From this comparison will frequently arise a *new* proportion, which may be requisite in solving a problem, or in carrying forward a demonstration. One of the most important cases is that in which two of the terms in one of the proportions compared, are the *same* with two in the other. The similar terms may be made to *disappear*, and a new proportion may be formed of the four remaining terms. For,

384. *If two ratios are respectively equal to a third, they are equal to each other.*[*]

This is nothing more than the 11th axiom applied to ratios.

1. If $\left. \begin{array}{l} a:b::m:n \\ \text{And } c:d::m:n \end{array} \right\}$ then $a:b::c:d$, or $a:c::b:d$. (Art.380.)

2. If $\left. \begin{array}{l} a:b::m:n \\ \text{And } m:n::c:d \end{array} \right\}$ then $a:b::c:d$, or $a:c::b:d$.

Cor. If $\left. \begin{array}{l} a:b::m:n \\ m:n \gt c:d \end{array} \right\}$ then $a:b \gt c:d$.[†]

For if the ratio of $m:n$ is greater than that of $c:d$, it is manifest that the ratio of $a:b$, which is *equal* to that of $m:n$, is also greater than that of $c:d$.

385. In these instances, the terms which are alike, in the two proportions, are the two *first* and the two *last*. But this arrangement is not essential. The order of the terms may be changed, in various ways, without affecting the equality of the ratios.

1. The similar terms may be the two *antecedents*, or the two *consequents*, in each proportion. Thus,

If $\left. \begin{array}{l} m:a::n:b \\ \text{And } m:c::n:d \end{array} \right\}$ then $\left\{ \begin{array}{l} \text{By alternation, } m:n::a:b \\ \text{And} \qquad\quad m:n::c:d. \end{array} \right.$

Therefore $a:b::c:d$, or $a:c::b:d$, by the last article,

Or if $\left. \begin{array}{l} a:m::b:n \\ \text{And } c:m::d:n \end{array} \right\}$ then $\left\{ \begin{array}{l} \text{By alternation, } a:b::m:n \\ \text{And} \qquad\quad c:d::m:n. \end{array} \right.$

Therefore $a:b::c:d$, or $a:c::b:d$.

* Euclid 11. 5.   † Euclid 12. 5.

A a

2. The *antecedents* in one of the proportions, may be the same as the *consequents* in the other.

If $\quad m:a::n:b$ } then { By inver. and altern. $a:b::m:n$
And $c:m::d:n$ { By alternation, $\quad c:d::m:n$

Therefore $a:b$, &c. as before.

Or if $a:m::b:n$ } then { By alternation, $\quad a:b::m:n$
And $\quad m:c::n:d$ { By inver. and altern. $c:d::m:n$

Therefore $a:b$, &c.

3. Two *homologous* terms, in one of the proportions, may be the same, as two *analogous* terms in the other.

If $\quad a:m::b:n$ } then { By alternation, $\quad a:b::m:n$
And $c:d::m:n$ { And $\qquad c:d::m:n$

Therefore $a:b$, &c.

Or if $a:b::m:n$ } then { The first is $\quad a:b::m:n$
And $\quad c:m::d:n$ { By alternation, $\quad c:d::m:n$

Therefore, &c.

All these are instances of an *equality*, between the ratios in one proportion, and those in another. In geometry, the proposition to which they belong is usually cited by the words "*ex aequo*," or "*ex aequali*."* The second case in this article is that which, in its form, most obviously answers to the explanation in Euclid. But they are all upon the same principle, and are frequently referred to, without discrimination.

386. Any number of proportions may be compared, in the same manner, if the two first or the two last terms in each preceding proportion, are the same with the two first or the two last in the following one.*

Thus if $a:b::c:d$ }
And $\quad c:d::h:l$ }
And $\quad h:l::m:n$ } then $a:b::x:y$.
And $\quad m:n::x:y$ }

That is, the two first terms of the first proportion have the same ratio, as the two last terms of the last proportion. For it is manifest that the ratio of *all* the couplets is the same.

And if the terms do not stand in the same order as here,

* Euclid 22. 5.

yet if they can be *reduced* to this form, the same principle is applicable.

Thus if
$$a:c::b:d$$
And $\quad c:h::d:l$
And $\quad h:m::l:n$
And $\quad m:x::n:y$
then by alternation
$$a:b::c:d$$
$$c:d::h:l$$
$$h:l::m:n$$
$$m:n::x:y$$
Therefore $a:b::x:y$, as before.

In all the examples in this, and the preceding articles, the two terms in one proportion which have equals in another, are neither the two *means*, nor the two *extremes*, but one of the means, and one of the extremes; and the resulting proportion is uniformly *direct*.

387. But if the two means, or the two extremes, in one proportion, be the same with the means, or the extremes, in another, the four remaining terms will be *reciprocally proportional*.

If $\quad a:m::n:b$
And $c:m::n:d$
then $a:c::\dfrac{1}{b}:\dfrac{1}{d}$, or $a:c::d:b$.

For $ab=mn$
And $cd=mn$
(Art. 374.) Therefore $ab=cd$, and $a:c::d:b$.

In this example, the two means in one proportion, are like those in the other. But the principle will be the same, if the *extremes* are alike, or if the extremes in one proportion are like the means in the other.

If $\quad m:a::b:n$
And $m:c::d:n$
then $a:c::d:b$.

Or if $a:m::n:b$
And $m:c::d:n$
then $a:c::d:b$.

The proposition in geometry which applies to this case is usually cited by the words "*ex aequo perturbate.*"*

388. Another way in which the terms of a proportion may be varied, is by *addition* or *subtraction*.

If to or from two analagous or two homologous terms of a proportion, two other quantities having the same ratio be added or subtracted, the proportion will be preserved.†

For a ratio is not altered, by adding to it, or subtracting from it, the terms of another equal ratio. (Art. 362.)

---

* Euclid 23. 5. † Euclid 2. 5.

$$\text{If} \quad a:b::c:d \bigg\}$$
$$\text{And} \quad a:b::m:n \bigg\}$$

Then by adding to, or subtracting from a and b, the terms of the equal ratio $m:n$, we have,

$$a+m:b+n::c:d \qquad \text{and } a-m:b-n::c:d.$$

And by adding and subtracting m and n, to and from c and d, we have,

$$a:b::c+m:d+n, \qquad \text{and } a:b::c-m:d-n.$$

Here the addition and subtraction are to and from *analogous* terms. But by alternation, (Art. 380.) these terms will become *homologous*, and we shall have,

$$a+m:c::b+n:d, \qquad \text{and } a-m:c::b-n:d.$$

Cor. 1. This addition may, evidently, be extended to *any number* of equal ratios.*

$$\text{Thus if } a:b:: \begin{cases} c:d \\ h:l \\ m:n \\ x:y \end{cases}$$

Then $a:b::c+h+m+x:d+l+n+y.$

Cor. 2. If $\begin{array}{l} a:b::c:d \\ \text{And } m:b::n:d \end{array} \bigg\}$ then $a+m:b::c+n:d.$†

For by alternation $\begin{array}{l} a:c::b:d \\ \text{And} \quad\quad m:n::b:d \end{array} \bigg\}$ there- $\bigg\{ \begin{array}{l} a+m:c+n::b:d \\ \text{fore } \text{or } a+m:b::c+n:d. \end{array}$

389. From the last article it is evident that if, in any proportion, the terms be added to, or subtracted from *each other*, that is,

If two analogous or homologous terms be added to, or subtracted from the two others, the proportion will be preserved.

Thus if $a:b::c:d,$ and $12:4::6:2,$ then,

1. *Adding* the two *last* terms, to the two *first*.

$a+c:b+d::a:b$	$12+6:4+2::12:4$
and $a+c:b+d::c:d$	$12+6:4+2::6:2$
or $a+c:a::b+d:b$	$12+6:12::4+2:4$
and $a+c:c::b+d:d.$	$12+6:6::4+2:2.$

* Euclid 2. 5. Cor.
† Euclid 24. 5.

2. *Adding* the two *antecedents*, to the two *consequents*.

$$a+b:b::c+d:d \qquad 12+4:4::6+2:2$$
$$a+b:a::c+d:c, \&c. \qquad 12+4:12::6+2:6, \&c.$$

This is called *Composition.**

3. *Subtracting* the two *first* terms, from the two *last*.

$$c-a:a::d-b:b$$
$$c-a:c::d-b:d, \&c.$$

4. *Subtracting* the two *last* terms, from the two *first*.

$$a-c:b-d::a:b\dagger$$
$$a-c:b-d::c:d, \&c.$$

5. *Subtracting* the *consequents*, from the *antecedents*.

$$a-b:b::c-d:d$$
$$a:a-b::c:c-d, \&c.$$

The alteration expressed by the last of these forms is called *Conversion*.

6. *Subtracting* the *antecedents*, from the *consequents*.

$$b-a:a::d-c:c$$
$$b:b-a::d:d-c, \&c.$$

7. Adding and subtracting.

$$a+b:a-b::a+d:a-d.$$

That is, the sum of the two first terms, is to their difference, as the sum of the two last, to their difference.

Cor. If any compound quantities, arranged as in the preceding examples, are proportional, the simple quantities of which they are compounded are proportional also.

Thus if $a+b:b::c+d:d$, then $a:b::c:d$.

This is called *Division.‡*

390. *If the corresponding terms of two or more ranks of proportional quantities be* MULTIPLIED *together, the products will be proportional.*

This is *compounding* ratios, (Art. 352.) or compounding proportions. It should be distinguished from what is called *composition*, which is an *addition* of the terms of a ratio. (Art. 389. 2.)

---

* Euclid 13. 5.   † Euclid 19. 5.

‡ Euclid 17. 5.   See Note II.

$$\left. \begin{array}{l} \text{If} \quad a:b::c:d \\ \text{And } h:l::m:n \end{array} \right\} \qquad \left. \begin{array}{l} 12:4::6:2 \\ 10:5::8:4 \end{array} \right\}$$

Then $\quad ah:bl::cm:dn.$ $\qquad 120:20::48:8.$

For, from the nature of proportion, the two ratios in the first rank are equal, and also the ratios in the second rank. And multiplying the corresponding terms is multiplying the *ratios*, (Art. 352. cor.) that is, multiplying *equals by equals*; (Ax. 3.) so that the ratios will still be equal, and therefore the four products must be proportional.

The same proof is applicable to any number of proportions.

$$\text{If} \left\{ \begin{array}{l} a:b::c:d \\ h:l::m:n \\ p:q::x:y \end{array} \right.$$

Then $ahp:blq::cmx:dny.$

From this it is evident, that if the terms of a proportion be multiplied, each into *itself*, that is, if they be *raised to any power*, they will still be proportional.

$$\begin{array}{ll} \text{If } a:b::c:d & \qquad 2:4::6:12 \\ a:b::c:d & \qquad 2:4::6:12 \end{array}$$

Then $a^2:b^2::c^2:d^2$ $\qquad\qquad 4:16::36:144$

Proportionals will also be obtained, by *reversing* this process, that is, by extracting the *roots* of the terms.

If $a:b::c:d,$ $\qquad$ then $\sqrt{a}:\sqrt{b}::\sqrt{c}:\sqrt{d}.$

For, taking the product of ext. and means, $ad=bc$
And extracting both sides, $\qquad \sqrt{ad}=\sqrt{bc}$
That is, (Arts. 259, 375.) $\qquad \sqrt{a}:\sqrt{b}::\sqrt{c}:\sqrt{d}.$

Hence,

391. If several quantities are proportional, *their like powers or like roots are proportional.*[*]

If $a:b::c:d$

-----

[*] It must not be inferred from this, that quantities have the same *ratio*, as their like powers or like roots. See art. 354.

Then $a^n : b^n :: c^n : d^n$. and $\sqrt[m]{a} : \sqrt[m]{b} :: \sqrt[m]{c} : \sqrt[m]{d}$.

And $\sqrt[m]{a^n} : \sqrt[m]{b^n} :: \sqrt[m]{c^n} : \sqrt[m]{d^n}$, that is, $a^{\frac{m}{n}} : b^{\frac{m}{n}} :: c^{\frac{m}{n}} : d^{\frac{m}{n}}$.

392. If the terms in one rank of proportionals be *divided* by the corresponding terms in another rank, the quotients will be proportional.

This is sometimes called the *resolution* of ratios.

$$
\begin{array}{ll}
\text{If} \quad a : b :: c : d \,\} & 12 : 6 :: 18 : 9 \,\} \\
\text{And } h : l :: m : n \,\} & 6 : 2 :: 9 : 3 \,\} \\
\text{Then } \dfrac{a}{h} : \dfrac{b}{l} :: \dfrac{c}{m} : \dfrac{d}{n} & \dfrac{12}{6} : \dfrac{6}{2} :: \dfrac{18}{9} : \dfrac{9}{3}
\end{array}
$$

This is merely *reversing* the process in art. 390, and may be demonstrated in a similar manner:

Or thus,

Taking the product of ext. and means, $\quad ad = bc \,\}$
And $\qquad\qquad\qquad\qquad\qquad\qquad\quad hn = lm \,\}$

Dividing one by the other, $\qquad\qquad\quad \dfrac{ad}{hn} = \dfrac{bc}{lm}$

That is, (Art. 375.) $\qquad\qquad\quad \dfrac{a}{h} : \dfrac{b}{l} :: \dfrac{c}{m} : \dfrac{d}{n}$.

This should be distinguished from what geometers call *division*, which is a *subtraction* of the terms of a ratio. (Art. 389. cor.)

When proportions are compounded by multiplication, it will often be the case, that the *same factor* will be found in two analogous or two homologous terms.

$$
\begin{array}{l}
\text{Thus if } a : b :: c : d \,\} \\
\text{And} \quad m : a :: n : c \,\} \\
\hline
am : ab :: cn : cd
\end{array}
$$

Here a is in the two first terms, and c in the two last. Dividing by these, (Art. 382.) the proportion becomes

$$m : b :: n : d. \quad \text{Hence,}$$

393. In compounding proportions, *equal factors*, or *divisors* in two analogous or homologous terms may be *rejected*.

$$\text{If } \begin{cases} a:b::c:d \\ b:h::d:l \\ h:m::l:n \end{cases} \qquad \begin{array}{l} 12:4::9:3 \\ 4:8::3:6 \\ 8:20::6:15 \end{array}$$

$$\text{Then } a:m::c:n. \qquad 12:20::9:15$$

This rule may be applied to the cases, to which the terms " *ex aequo,*" and " *ex aequo perturbate*" refer. See arts. 385 and 387. One of the methods may serve to verify the other.

394. The changes which may be made in proportions, without disturbing the equality of the ratios, are so numerous, that they would become burdensome to the memory, if they were not reducible to a few general principles. They are mostly produced,

1. By inverting the *order* of the terms, Art. 380.
2. By *mult.* or *dividing* by the *same quantity,* Art. 382.
3. By compar. propor's which have *like terms,* Art. 384,5,6,7.
4. By *add.* or *subt.* the terms of equal ratios, Art. 388,9.
5. By *mult.* or *divid.* one propor. by another, Art. 390,2,3.
6. By *involv.* or *extract. the roots* of the terms, Art. 391.

395. When four quantities are proportional, if the *first* be greater than the *second,* the *third* will bo greater than the *fourth ;* if equal, equal ; if less, less.

For, the ratios of the two couplets being the same, if one is a ratio of *equality,* the other is also, and therefore the antecedent in each is *equal* to its consequent ; (Art. 350.) if one is a ratio of *greater inequality,* the other is also, and therefore the antecedent in each is *greater* than its consequent ; and if one is a ratio of *less inequality,* the other is also, and therefore the antecedent in each is *less* than its consequent.

Let $a:b::c:d;$ then if $\begin{cases} a=b, & c=d \\ a>b, & c>d \\ a<b, & c<d. \end{cases}$

Cor. 1. If the *first* be greater than the *third,* the *second* will be greater than the fourth ; if equal, equal ; if less, less.*

* Euclid 14. 5.

For by alternation $a:b::c:d$ becomes $a:c::b:d$, without any alteration of the quantities. Therefore if $a=b, c=d$, &c. as before.

Cor. 2. If $\left.\begin{array}{l} a:m::c:n \\ \text{And } m:b::n:d \end{array}\right\}$ then if $a=b, c=d$, &c.*

For, by equality of ratios, (Art. 385. 2.) or compounding ratios, (Arts. 390. 393.)

$a:b::c:d$. Therefore, if $a=b, c=d$, &c. as before.

Cor. 3. If $\left.\begin{array}{l} a:m::n:d \\ \text{And } m:b::c:n \end{array}\right\}$ then if $a=b, c=d$, &c.†

For, by compounding ratios, (Arts. 390, 393.)

$a:b::c:d$. Therefore, if $a=b, c=d$, &c.

CONTINUED PROPORTION.

396. When quantities are in continued proportion, *all* the ratios are *equal*. (Art. 372.) If

$$a:b::b:c::c:d::d:e,$$

the ratio of $a:b$ is the same, as that of $b:c$, of $c:d$, or of $d:e$. The ratio of the *first* of these quantities to the *last*, is equal to the *product* of all the intervening ratios; (Art. 353.) that is, the ratio of $a:e$ is equal to

$$\frac{a}{b} \times \frac{b}{c} \times \frac{c}{d} \times \frac{d}{e}.$$

But as the intervening ratios are all *equal*, instead of multiplying them into each other, we may multiply any one of them into *itself*; observing to make the number of factors equal to the number of intervening ratios. Thus the ratio of $a:e$, in the example just given, is equal to

$$\frac{a}{b} \times \frac{a}{b} \times \frac{a}{b} \times \frac{a}{b} = \frac{a^4}{b^4}.$$

When several quantities are in continued proportion, the number of couplets, and of course, the number of ratios, is *one less* than the number of quantities. Thus the five proportional quantities a, b, c, d, e, form four couplets contain-

---

* Euclid 20. 5.   † Euclid 21. 5.

Bb

ing four ratios; and the ratio of $a:c$ is equal to the ratio of $a^4:b^4$, that is, the ratio of the fourth power of the first quantity, to the fourth power of the second. Hence,

397. If *three* quantities are proportional, *the first is to the third, as the square of the first, to the square of the second;* or as the square of the second, to the square of the third. In other words, the first has to the third, a *duplicate* ratio of the first to the second.

If $a:b::b:c$, then $a:c::a^2:b^2$. And universally,

398. If several quantities are proportional, the ratio of the first to the last is equal to one of the intervening ratios raised to a power whose index is one less than the number of quantities.

If there are *four* proportionals a, b, c, d, then $a:d::a^3:b^3$.
If there are *five* a, b, c, d, e, $a:e::a^4:b^4$,&c.

399. If several quantities are proportional, they will be proportional when the order of the whole is *inverted.* This has already been proved, with respect to *four* proportional quantities. (Art. 380. cor.) It may be extended to any number of quantities.

Between the numbers,   64, 32, 16, 8, 4,
The ratios are   2, 2, 2, 2,
Between the same inverted   4, 8, 16, 32, 64,
The ratios are $\frac{1}{2}$, $\frac{1}{2}$, $\frac{1}{2}$, $\frac{1}{2}$.

So if the order of any proportional quantities be inverted, the ratios in one series will be the *reciprocals* of those in the other. For, by the inversion, each antecedent becomes a consequent, and *v. v.* and the ratio of a consequent to its antecedent is the reciprocal of the ratio of the antecedent to the consequent. (Art 351.) That the reciprocals of equal quantities are themselves equal, is evident from ax. 4.

400. HARMONICAL OR MUSICAL PROPORTION may be considered as a species of geometrical proportion. It consists in an equality of geometrical ratios; but one or more of the terms is the *difference* between two quantities.

Three or *four* quantities are said to be in *harmonical proportion,* when the *first* is to the *last,* as the difference be-

.tween the *two first*, to the difference between the *two last*.

If the *three* quantities a, b, and c, are in harmonical proportion, then $a:c::a-b:b-c$.

If the *four* quantities a, b, c, and d, are in harmonical proportion, then $a:d::a-b:c-d$.

Thus the three numbers 12, 8, 6, are in harmonical proportion.

And the four numbers 20, 16, 12, 10, are in harmonical proportion.

401. If, of four quantities in harmonical proportion, any three be given, the other may be found. For, from the proportion

$$a:d::a-b:c-d,$$

by taking the product of the extremes and the means, we have $ac-ad=ad-bd$.

And this equation may be reduced, so as to give the value of either of the four letters.

Thus by transposing $-ad$, and dividing by a,

$$c=\frac{2ad-bd}{a}.$$

402. A list of the articles in this section which contain the propositions in the 5th book of Euclid.*

Prop.	I.	Art. 363.		XIII.	384, cor.
	II.	388.		XIV.	395, cor. 1.
	III.	382.		XV.	360.
	IV.	382, cor. 1.		XVI.	380.
	V.	362.		XVII.	389, cor.
	VI.	362.		XVIII.	389, 2.
	VII.	349, cor. 1.		XIX.	389, 4.
	VIII.	357, cor. 358, cor.		XX.	395, cor. 2.
	IX.	349, cor. 2.		XXI.	395, cor. 3.
	X.	357, cor. 358, cor.		XXII.	386.
	XI.	384.		XXIII.	387.
	XII.	363.		XXIV.	388, cor. 2.

* See Note I.

VARIATION or GENERAL PROPORTION.[*]

ART. 403. THE quantities which constitute the terms of a proportion are, frequently, so related to each other, that, if one of them be either increased or diminished, another depending on it will also be increased or diminished, in such a manner, that the proportion will still be preserved. If the value of 50 yards of cloth is 100 dollars, and the quantity be reduced to 40 yards; the value will, of course, be reduced to 80 dollars: if the quantity be reduced to 30 yards, the value will be reduced to 60 dollars, &c.

$$yd. \quad yd. \quad dol. \quad dol.$$

That is $50 : 40 :: 100 : 80$

$50 : 30 :: 100 : 60$

$50 : 20 :: 100 : 40$, &c.

As the consequent of the *first* couplet is varied, the consequent of the *second* is varied, in such a manner, that the proportion is constantly preserved.

If the two antecedents are A and B; and if a represents a quantity of the *same kind* with A, but either greater or less; and b, a quantity of the same kind with B, but as many times greater or less, as a is greater or less than A; then

$$A : a :: B : b;$$

that is, if A by varying becomes a, then B becomes b. This is expressed more concisely, by saying that A *varies as* B, or A *is as* B. Thus the *wages* of a labouring man vary as the *time* of his service. We say that the *interest* of money which is loaned for a given time, is *proportioned* to the principal. But a proportion contains *four terms*. Here are only two, the *interest* and the *principal*. This then is an *abridged statement*, in which two terms are mentioned instead of four. The proportion in form would be;

[*] Newton's Princip. Book I. Sec. 1. Lemma 10, schol. Emerson on Proportion, Wood's Algebra, Lydlam's Math. Saunderson's Algebra, Art. 299.

As any given principal, is to any other principal;
So is the interest of the former, to the interest of the latter.

$$P. \quad P. \quad In. \quad In.$$

For instance, 100 : 300 :: 6 : 18.

404. In many mathematical and philosophical investigations, we have occasion to determine the general relations of certain classes of quantities to each other, without limiting the inquiry to any particular values of those quantities. In such cases, it is frequently sufficient to mention only two of the terms of a proportion. It must be kept in mind, however, that four are always *implied*. When it is said, for instance, that the weight of water is proportioned to its bulk, we are to understand;

That *one* gallon, is to any *number* of gallons;
As the *weight* of one gallon, is to the weight of the given number of gallons.

405. The character $\propto$ is used to express the proportion of variable quantities.

Thus $A \propto B$ signifies that A *varies* as B, that is, that

$$A : a :: B : b.$$

The expression $A \propto B$ may be called a *general proportion*.

406. One quantity is said to vary *directly* as another, when the one increases as the other increases, or is diminished as the other is diminished, so that

$$A \propto B, \text{ that is } A : a :: B : b.$$

The interest on a loan is increased or diminished, in proportion to the principal. If the principal is doubled, the interest is doubled; if the principal is trebled, the interest is trebled, &c.

407. One quantity is said to vary *inversely* or *reciprocally* as another, when the one is proportioned to the reciprocal of the other; that is, when the one is diminished, as the other is increased, so that

$$A \propto \frac{1}{B}, \text{ that is, } A : a :: \frac{1}{B} : \frac{1}{b}, \text{ or } A : a :: b : B.$$

In this case, if A is greater than a, B is less than b. (Art. 395.) The time required for a man to raise a given sum, by his labour, is inversely as his wages. The higher his wages, the less the time.

408. One quantity is said to vary as *two others jointly*, when

the one is increased or diminished, as the *product* of the
other two, so that

$$A \propto BC, \text{ that is } A : a :: BC : bc.$$

The interest of money varies as the product of the principal and time. If the time be doubled, and the principal doubled, the interest will be four times as great.

409. One quantity is said to vary *directly* as a *second*, and *inversely* as a *third*, when the first is always proportioned to the second divided by the third, so that

$$A \propto \frac{B}{C}, \text{ that is } A : a :: \frac{B}{C} : \frac{b}{c}.$$

410. To understand the methods by which the statements of the relations of variable quantities are changed from one form to another, little more is necessary, than to make an *application* of the principles of common proportion; bearing constantly in mind, that a general proportion is only an abridged expression, in which two terms are mentioned instead of four. When the deficient terms are supplied, the reason of the several operations will commonly be apparent.

411. It is evident, in the first place, that the *order of the terms* in a general proportion may be *inverted*, (Art. 369.)

If $A : a :: B : b$, that is, if $A \propto B$;
Then $B : b :: A : a$, that is, $B \propto A$.

412. If one or both the terms in a general proportion, be *multiplied* or *divided* by a constant quantity, the proportion will be preserved.

For multiplying or dividing one or both of the terms is the same, as multiplying or dividing *analogous* terms in the proportion expressed at length. (Art. 382. and cor. 1.)

If $A : a :: B : b$, that is, if $A \propto B$,
Then $mA : ma :: B : b$, that is, $mA \propto B$,
And $mA : ma :: mB : mb$, that is, $mA \propto mB$. &c.

413. If *both* the terms be multiplied or divided even by a *variable* quantity, the proportion will be preserved. For this is equivalent to multiplying the two *antecedents* by one quantity, and the two *consequents* by another. (Art. 382.)

If $A : a :: B : b$, that is, if $A \propto B$;
Then $MA : ma :: MB : mb$, that is, $MA \propto MB$. &c.

Cor. 1. If one quantity varies as another, the *quotient of* the one divided by the other is *constant.* In other words, if the numerator of a fraction varies as the denominator, the *value* remains the same.

If $A:a::B:b$, that is, if $A \propto B$,

$$\text{Then } \frac{A}{B}:\frac{a}{b}::\frac{B}{B}:\frac{b}{b}::1:1. \quad (\text{Art. 128.})$$

Here the third and fourth terms are equal, because each is equal to 1. Of course the two first terms are equal; (Art. 395.) so that, if A be increased or diminished as many times as B, the *quotient* will be invariably the same.

Cor. 2. If the *product* of two quantities is *constant*, one varies *reciprocally* as the other.

$$\text{If } AB:ab::1:1, \text{ then } \frac{AB}{B}:\frac{ab}{b}::\frac{1}{B}:\frac{1}{b}, \text{ or } A:a::\frac{1}{B}:\frac{1}{b}$$

Cor. 3. Any *factor* in one term of a general proportion, may be *transferred*, so as to become a *divisor* in the other; and v. v.

If $A \propto BC$, then dividing by B, $\frac{A}{B} \propto C$. (Art. 118.)

If $A \propto \frac{1}{CD}$, then mult. by C, $AC = \frac{1}{D}$. (Art. 159.)

414. If two quantities vary respectively as a third, then one of the two varies as the other. (Art. 384.)

$$\left.\begin{array}{l} \text{If} \quad A:a::B:b \\ \text{And} \quad C:c::B:b \end{array}\right\} \text{ that is, if } \left\{\begin{array}{l} A \propto B \\ C \propto B; \end{array}\right.$$
$$\text{Then} \quad A:a::C:c, \text{ that is, } \quad A \propto C.$$

415. If two quantities vary respectively as a third, their *sum* or *difference* will vary in the same manner. (Art. 388.)

$$\left.\begin{array}{l} \text{If} \quad A:a::B:b \\ \text{And} \quad C:c::B:b \end{array}\right\} \text{ that is, if } \left\{\begin{array}{l} A \propto B \\ C \propto B; \end{array}\right.$$
$$\text{Then } A+C:a+c::B:b, \text{ that is, } A+C \propto B,$$
$$\text{And } A-C:a-c::B:b, \text{ that is, } A-C \propto B.$$

Cor. The addition here may be extended to *any number* of quantities all varying alike. (Art. 388. cor. 1.)
If $A \propto B$, and $C \propto B$, and $D \propto B$, and $E \propto B$, then
$$(A+C+D+E) \propto B.$$

416. The terms of one general proportion may be multiplied or divided by the corresponding terms of another. (Art. 390.)

$$\left. \begin{array}{l} \text{If } A:a::B:b \\ \text{And } C:c::D:d \end{array} \right\} \text{ that is, if } \left\{ \begin{array}{l} A \propto B \\ C \propto D; \end{array} \right.$$

Then $AC:ac::BD:bd$ that is, $AC \propto BD$.

Cor. If two quantities vary respectively as a third, the product of the two will vary as the *square* of the other.

$$\left. \begin{array}{l} \text{If } A \propto B \\ \text{And } C \propto B \end{array} \right\} \text{ then } AC \propto B^2.$$

417. If any quantity vary as another, any *power* or *root* of the former will vary, as a like power or root of the latter. (Art. 391.)

If $A:a::B:b,$ that is, if $A \propto B,$

Then $A^n:a^n::B^n:b^n.$ that is; $A^n \propto B^n,$

And $A^{\frac{1}{n}}:a^{\frac{1}{n}}::B^{\frac{1}{n}}:b^{\frac{1}{n}},$ that is, $A^{\frac{1}{n}} \propto B^{\frac{1}{n}}.$

418. In compounding general proportions, equal *factors* or *divisors*, in the two terms, may be rejected. (Art. 393.)

$$\left. \begin{array}{l} \text{If } A:a::B:b \\ \text{And } B:b::C:c \\ \text{And } C:c::D:d \end{array} \right\} \text{ that is, if } \left\{ \begin{array}{l} A \propto B \\ B \propto C \\ C \propto D \end{array} \right.$$

Then $A:a::D:d,$ that is, $A \propto D$

Cor. If one quantity varies as a second, the second, as a third, the third, as a fourth, &c. then the *first* varies as the last.

If $A \propto B \propto C \propto D,$ then $A \propto D.$

If $A \propto B \propto \dfrac{1}{C},$ then $A \propto \dfrac{1}{C}$; that is, if the first varies *directly* as the second, and the second varies *reciprocally* as the third; the first varies reciprocally as the third.

419. If any quantity vary as the *product* of two others, and if one of the latter be considered *constant*, the first will vary as the other.

If $W \propto LB$, and if B be constant, then $W \propto L.$

Here it must be observed, that there are *two conditions*; First, that W varies as the *product* of the two other quantities;

Secondly, that one of these quantities B is *constant*.

Then, by the conditions, $W:w::LB:lB$; B being the same in both terms.

Divid. by the constant quantity B, $W:w::L:l$, that is $W \propto L$.

And if L be considered constant, $\qquad\qquad W \propto B$.

Thus the weight of a board, of uniform thickness and density, varies as its length and breadth. If the *length* is given, the weight varies as the breadth. And if the *breadth* is given, the weight varies as the length.

Cor. The same principle may be extended to any number of quantities. The weight of a stick of timber, of given density, depends on the length, breadth, and thickness. If the length is given, the weight varies as the breadth and thickness. If the length and breadth are given, the weight varies as the thickness, &c.

If $\qquad\qquad\qquad\qquad\qquad\qquad W \propto LBT$;

Then making L constant, $\qquad\qquad\quad W \propto BT$;

And making L and B constant, $\qquad\quad W \propto T$.

420. On the other hand, if one quantity depends on two others; so that when the second is given, the first varies as the third, and when the third is given, the first varies as the second; then the first varies as the *product* of the other two.

If the weight of a board varies as the length, when the breadth is given, and as the breadth when the length is given; then if the length and breadth *both* vary, the weight varies as their product.

If $\quad W \propto L$, when B is constant, $\Big\}$ then $W \propto BL$.
And $W \propto B$, when L is constant, $\Big\}$

In demonstrating this, we have to consider, *two variable values* of W; one, when L *only* varies, and the other, when L and B *both* vary.

Let $w'=$ the first of these variable values,
And $w=$ the other;

So that W will be changed to w', by the varying of L;
And w' will be farther changed to w, by the varying of B.

Then, by the supposition, $W:w'::L:l$, when B is constant.
And $\qquad\qquad\qquad\qquad w':w::B:b$, when B varies.

Mult. correspond. terms, $Ww':ww'::BL:bl$. (Art. 390.)
Divid. by w' (Art. 382.) $\quad W:w::BL:bl$, i. e. $W \propto BL$.

The proof may be extended to any number of quantities. The weight of a piece of timber, depends on its length, breadth, thickness and density. If any three of these are given, the weight varies as the other.

This case must not be confounded with that in art. 416, cor. In that, B is supposed to vary as A and as C, *at the same time*. In this, B varies as A, only when C is constant, and as C, only when A is constant. It can not therefore vary as A and as C separately, at the same time.

421. Many writers, in expressing a general proportion, do not use the term *vary*, or the character which has here been put for it. Instead of $A \propto B$, they say simply that A *is as B*. See Enfield's Philosophy. It may be proper to observe also, that the word *given* is frequently used to distinguish *constant* quantities, from those which are variable; as well as to distinguish *known* quantities, from those which are unknown. (Art. 17.)

ARITHMETICAL and GEOMETRICAL PROGRESSION.

ART. 422. QUANTITIES which decrease by a common difference, as the numbers 10, 8, 6, 4, 2, are in continued arithmetical proportion. (Art. 372.) Such a series is also called a *progression*, which is only another name for continued proportion.

It is evident, that the proportion will not be destroyed, if the order of the quantities be *inverted.* Thus the numbers 2, 4, 6, 8, 10, are in continued arithmetical proportion.

Quantities, then, *are in arithmetical progression, when they increase or decrease by a common difference.*

When they *increase,* they form what is called an *ascending* series, as 3, 5, 7, 9, 11, &c.

When they *decrease,* they form a *descending* series, as 11, 9, 7, 5, &c.

The natural numbers 1, 2, 3, 4, 5, 6, &c. are in arithmetical progression ascending.

423. From the definition it is evident that, in an ascending series, each succeeding term is found, by *adding the common difference* to the preceding term.

If the first term is 3, and the common difference 2;

Then $3+2=5$ is the second term, $7+2=9$ the fourth.

$\qquad 5+2=7$ the third, $\qquad 9+2=11$ the fifth, &c.

And the series is 3, 5, 7, 9, 11, 13, &c.

If the first term is a, and the common difference d;

Then $a+d$ is the second term, $a+2d+d=a+3d$ the fourth,

$\qquad a+d+d=a+2d$ the 3d, $a+3d+d=a+4d$ the 5th,&c.

And the series is $\overset{1}{a}, \overset{2}{a+d}, \overset{3}{a+2d}, \overset{4}{a+3d}, \overset{5}{a+4d}$, &c.

If the first term and the common difference are the *same,*

the series becomes more simple. Thus if a is the first term, and the common difference, and n the number of terms. :

Then $a+a=2a$ is the second term
$2a+a=3a$ the third, &c.

And the series is a, $2a$, $3a$, $4a$ ..... na.

424. In a *descending* series, each succeeding term is found, by *subtracting the common difference* from the preceding term.

If a is the first term, and d the common difference, the series is $\overset{1}{a}$, $\overset{2}{a-d}$, $\overset{3}{a-2d}$, $\overset{4}{a-3d}$, $\overset{5}{a-4d}$, &c.

In this manner, we may obtain any term, by continued addition or subtraction. But in a long series, this process would become tedious. There is a method much more expeditious. By attending to the series

$$\overset{1}{a}, \overset{2}{a+d}, \overset{3}{a+2d}, \overset{4}{a+3d}, \overset{5}{a+4d}, \&c.$$

it will be seen, that the number of times d is added to a is *one less* than the number of the term.

The *second* term is $a+d$, i. e. a added to *once* d;
The *third* is $a+2d$, a added to *twice* d;
The *fourth* is $a+3d$, a added to *thrice* d, &c.

So, if the series be continued,

The 50th term will be $a+49d$
The 100th term $a+99d$

In the *greatest* term, the number of times d is added to a, is *one less* than the number of *all* the terms. If then $a=$ the least term, $z=$ the greatest, $n=$ the number of terms, we shall have, in all cases, $z=a+(n-1)\times d$; that is,

425. In an arithmetical progression, *the greatest term is equal to the least*, $+$ *the product of the common difference into the number of terms less one.*

Any other term may be found in the same way. For the series may be made to stop at any term, and that may be considered, for the time, as the last.

Thus the mth term $=a+(m-1)\times d$.

If the first term and the common difference are the *same*,
$$z=a+(n-1)a=a+na-a, \text{ that is, } z=na.$$

In an *ascending* series, the first term is, evidently, the least, and the last, the greatest. But in a descending series, the first term is the greatest, and the last, the least.

426. The equation $z=a+(n-1)d$ not only shows the value of the greatest term, but, by a few simple reductions, will enable us to find other parts of the series. It contains four different quantities,

a, the *least* term, n, the *number* of terms, and
z, the *greatest* term, d, the *common difference*.

If any three of these be given, the other may be found.

1. By the equation already found,
$$z=a+(n-1)d=\text{\textit{the greatest term.}}$$

2. Transposing $(n-1)d$, (Art. 173.)
$$z-(n-1)d=a=\text{\textit{the least term.}}$$

3. Transposing a in the 1st and dividing by $n-1$,
$$\frac{z-a}{n-1}=d=\text{\textit{the common difference.}}$$

4. Transp. a in the 1st, dividing by d, and transp. -1,
$$\frac{z-a}{d}+1=n=\text{\textit{the number of terms.}}$$

Prob. 1. If the first term of an increasing progression is 7, the common difference 3, and the number of terms 9, what is the last term?
Ans. $z=a+(n-1)d=7+(9-1)\times3=31.$
And the series is 7, 10, 13, 16, 19, 22, 25, 28, 31.

Prob. 2. If the last term of an increasing progression is 60, the number of terms 12, and the common difference 5, what is the first term?
Ans. $a=z-(n-1)d=60-(12-1)\times5=5.$

427. There is one other inquiry to be made concerning a series in arithmetical progression. It is often necessary to find the *sum of all the terms*. This is called the *summation* of the series. The most obvious mode of obtaining the amount of the terms, is to add them together. But the nature of progression will furnish us with a method more expeditious.
It is manifest that the sum of the terms will be the same,

in whatever *order* they are written. The sum of the ascending series 3, 5, 7, 9, 11, is the same, as that of the descending series 11, 9, 7, 5, 3. The sum of *both* the series is, therefore, *twice* as great, as the sum of the terms in one of them. There is an easy method of finding this *double sum*, and, of course, the sum itself which is the object of inquiry. Let a given series be written, both in the direct, and in the inverted order, and then add the corresponding terms together.

Take, for instance, the series   3, 5, 7, 9, 11,
And the same inverted   11, 9, 7, 5, 3.

The sums of the terms will be   14, 14, 14, 14, 14.

Take also the series a, $a+d$, $a+2d$, $a+3d$, $a+4d$,
And the same inver. $a+4d$, $a+3d$, $a+2d$, $a+d$, a.

The sums will be $2a+4d, 2a+4d, 2a+4d, 2a+4d, 2a+4d$.

Here we discover the important property, that,

428. In an arithmetical progression, *the sum of the extremes is equal to the sum of any other two terms equally distant from the extremes.*

In the series of numbers above, the sum of the first and the last term, of the first but one and the last but one, &c. is 14. And in the other series, the sum of each pair of corresponding terms is $2a+4d$.

To find the sum of *all* the terms in the double series, we have only to observe, that it is equal to the sum of the extremes repeated as many times as there are terms.

The sum of 14, 14, 14, 14, $14 = 14 \times 5$.

And the sum of the terms in the other double series is $(2a+4d) \times 5$.

But this is *twice* the sum of the terms in the single series. If then we put

a = the least term, n = the number of terms,
z = the greatest, s = the sum of the terms.

we shall have this equation,

$$s = \frac{a+z}{2} \times n. \quad \text{That is,}$$

429. In an arithmetical progression. the *sum of all the terms is equal to half the sum of the extremes multiplied into the number of terms.*

Prob. What is the sum of the natural series of numbers 1, 2, 3, 4, 5, &c. up to 1000?

$$\text{Ans. } s = \frac{a+z}{2} \times n = \frac{1+1000}{2} \times 1000 = 500500.$$

430. In the series of *odd* numbers 1, 3, 5, 7, 9, &c. continued to any given extent, the last term is always one less than twice the number of terms.

For $z = a + (n-1)d$. (Art. 425.) But in the proposed series $a = 1$, and $d = 2$.

The equation, then, becomes $z = 1 + (n-1) \times 2 = 2n - 1$.

431. In the series of odd numbers 1, 3, 5, 7, 9, &c. *the sum of the terms is always equal to the square of the number of terms.*

$$\text{For } s = \tfrac{1}{2}(a+z)n. \text{ (Art. 429.)}$$

But here $a = 1$, and by the last article, $z = 2n - 1$.

The equation, then becomes, $s = \tfrac{1}{2}(1 + 2n - 1)n = n^2$.

$$\left.\begin{array}{l} \text{Thus } 1+3=4 \\ 1+3+5=9 \\ 1+3+5+7=16 \end{array}\right\} \text{the square of the number of terms.}$$

Or thus,

Series of numbers,	1, 3, 5, 7, 9, 11, 13, 15, &c.
Number of terms,	1, 2, 3, 4, 5, 6, 7, 8, &c.
Sum of the terms,	1, 4, 9, 16, 25, 36, 49, 64. &c.

432. If there be two ranks of quantities in arithmetical progression, the *sums* or *differences* will also be in arithmetical progression.

For, by the addition or subtraction of the corresponding terms, the *ratios* are added or subtracted. (Art. 345.) And by the nature of progression, all the ratios in the series are *equal*. Therefore equal ratios being added to, or subtracted from, equal ratios, the new ratios thence arising will also be equal.

$$\left.\begin{array}{lccccccc} \text{To and from} & 3, & 6, & 9, 12, 15, 18, 21 \\ \text{Add and sub.} & 2, & 4, & 6, \ 8, 10, 12, 14 \\ \hline \text{Sums} & 5, & 10, & 15, 20, 25, 30, 35 \\ \text{Diff.} & 1, & 2, & 3, \ 4, \ 5, \ 6, \ 7 \end{array}\right\} \text{whose ratio is} \left\{\begin{array}{l} 3 \\ 2 \\ - \\ 5 \\ 1 \end{array}\right.$$

433. If all the terms of an arithmetical progression be

multiplied or *divided* by the same quantity, the products or quotients will be in arithmetical progression.

For, by the multiplication or division of the terms, the *ratios* are multiplied or divided; (Art. 344.) that is, equal quantities are multiplied or divided by the given quantity. They will therefore remain equal.

If the series   3, 5, 7, 9,11,&c. be mult. by 4;
The prods. will be 12,20,28,36,44,&c. and if this be divid. by 2;
The quots. will be   6,10,14,18,22,&c.

GEOMETRICAL PROGRESSION.

434. As arithmetical proportion continued is arithmetical progression, so geometrical proportion continued is geometrical progression.

The numbers 64, 32, 16, 8, 4, are in continued geometrical proportion. (Art. 372.)

In this series, if each preceding term be *divided* by the common ratio, the quotient will be the following term.

$$\tfrac{64}{2}=32, \text{ and } \tfrac{32}{2}=16, \text{ and } \tfrac{16}{2}=8, \text{ and } \tfrac{8}{2}=4.$$

If the order of the series be *inverted*, the proportion will still be preserved; (Art. 399.) and the common divisor will become a multiplier.   In the series

4,8,16,32,64,&c. $4 \times 2=8$, and $8 \times 2=16$, and $16 \times 2=32$,&c.

435. *Quantities, then, are in geometrical progression, when they increase by a common multiplier, or decrease by a common divisor.*

The common multiplier or divisor is called the *ratio.* In a *descending* series, it is, as in common proportion, a *direct* ratio. But in an *ascending* series, it is a *reciprocal* ratio.

In the series 4,8,16,32,&c. $\tfrac{8}{4}=\tfrac{16}{8}=\tfrac{32}{16}=2$,the common ratio.

But if the direct ratios are equal, the reciprocal ratios are also equal. (Art. 399.) So that quantities in geometrical progression, whether ascending or descending, may be considered proportionals.

To investigate the properties of geometrical progression, we may take nearly the same course, as in arithmetical progression, observing to substitute continual *multiplication and division,* instead of addition and subtraction. It is evident, in the first place, that,

436. In an ascending geometrical series, each succeeding term is found by *multiplying the ratio* into the preceding term.

If the first term is a, and the ratio r,

Then $a \times r = ar$, the second term, $ar^2 \times r = ar^3$, the fourth,
$ar \times r = ar^2$, the third, $ar^3 \times r = ar^4$, the fifth, &c.

And the series is a, ar, ar^2, ar^3, ar^4, ar^5, &c.

437. If the first term and the ratio are the *same*, the progression is simply a series of powers.

If the first term and the ratio are each equal to r,

Then $r \times r = r^2$, the second term, $r^3 \times r = r^4$, the fourth,
$r^2 \times r = r^3$, the third , $r^4 \times r = r^5$, the fifth.

And the series is r, r^2, r^3, r^4, r^5, r^6, &c.

438. In a *descending* series, each succeeding term is found by *dividing* the preceding term by the ratio.

If the first term is ar^6, and the ratio r,

The series is ar^6, ar^5, ar^4, ar^3, ar^2, ar, a, &c.

If the first term is a and the ratio r,

The series is a, $\dfrac{a}{r}$, $\dfrac{a}{r^2}$, $\dfrac{a}{r^3}$, &c, or (Art.207.) a, a^{r-1}, a^{r-2}, &c.

If the first term is 1, and the ratio 2,

The series is 1, $\frac{1}{2}$, $\frac{1}{4}$, $\frac{1}{8}$, $\frac{1}{16}$, $\frac{1}{32}$, $\frac{1}{64}$, &c.

By attending to the series $\overset{1}{a}$, $\overset{2}{ar}$, $\overset{3}{ar^2}$, $\overset{4}{ar^3}$, $\overset{5}{ar^4}$, $\overset{6}{ar^5}$, &c. it will be seen that, in each term, the exponent of the power of the ratio is *one less*, than the number of the term.

If then $a =$ the least term, $r =$ the ratio
$z =$ the greatest, $n =$ the number of terms;
we have the equation $z = ar^{n-1}$, that is,

439. In geometrical progression, *the greatest term is equal to the product of the least, into that power of the ratio whose index is one less than the number of terms.*

When the least term and the ratio are the *same*, the equation becomes $z = rr^{n-1} = r^n$. See art. 437.

440. Of the four quantities a, z, r, and n, any *three* being given, the other may be found.*

* See Note K.

Dd

1. By the last article,

$$z = ar^{n-1} = \text{the } greatest \text{ term.}$$

2. Dividing by r^{n-1},

$$\frac{z}{r^{n-1}} = a = \text{the } least \text{ term.}$$

3. Divid. the 1st by a, and extracting the root, (Art. 297.)

$$\left(\frac{z}{a}\right)^{\frac{1}{n-1}} = r = \text{the } ratio.$$

441. The next thing to be attended to is the rule for finding the *sum of all the terms*.

If any term, in an increasing geometrical series, be multiplied by the ratio, the product will be the succeeding term. (Art. 436.) Of course, if *each* of the terms be multiplied by the ratio, a new series will be produced, in which all the terms except the last will be the same, as all except the first in the other series. To make this plain, let the new series be written under the other, in such a manner, that each term shall be removed one step to the right of that from which it is produced in the line above.

Take, for instance, the series   2, 4, 8, 16, 32
Mult. each term by the ratio, we have   4, 8, 16, 32, 64.

Here it will be seen, at once, that the four last terms in the upper line are the same, as the four first in the lower line. The only terms which are not in *both*, are the *first* of the one series, and the *last* of the other. So that when we subtract the one series from the other, all the terms except these two will disappear, by balancing each other.

If the given series is $a, ar, ar^2, ar^3, \ldots ar^{n-1}.$
Then mult. by r, we have, $ar, ar^2, ar^3, \ldots ar^{n-1}, ar^n.$

Now let $s =$ the sum of the terms,

Then $s = a + ar + ar^2 + ar^3, \ldots + ar^{n-1},$
And mult. by r, $rs = $ $ar + ar^2 + ar^3, \ldots + ar^{n-1} + ar^n.$

Subt'g the first equation from the second, $rs - s = ar^n - a$

And dividing by $(r-1)$, (Art. 121.) $\qquad s = \dfrac{ar^n - a}{r - 1}.$

In this equation, ar^n is the last term in the new series, and is therefore the product of the ratio into the last term in the *given* series. Hence,

442. The sum of a series in geometrical progression is found, by multiplying the greatest term into the ratio, subtracting the least term, and dividing the remainder by the ratio less one.

Prob. If in a series of numbers in geometrical progression, the first term is 6, the last term 1458, and the ratio 3, what is the sum of all the terms?

$$\text{Ans. } s = \frac{rz - a}{r - 1} = \frac{3 \times 1458 - 6}{3 - 1} = 2184.$$

443. *Quantities in geometrical progression are proportional to their differences.*

Let the series be a, ar, ar^2, ar^3, ar^4, &c.

By the nature of geometrical progression,

$$a : ar :: ar : ar^2 :: ar^2 : ar^3 :: ar^3 : ar^4, \&c.$$

In each couplet let the antecedent be subtracted from the consequent according to art. 389. 6.

Then $a : ar :: ar - a : ar^2 - ar :: ar^2 - ar : ar^3 - ar^2$, &c.

That is, the first term is to the second, as the difference between the first and second, to the difference between the second and third; and as the difference between the second and third, to the difference between the third and fourth, &c.

Cor. If quantities are in geometrical progression, their *differences* are also in geometrical progression.

Thus the numbers   3, 9, 27, 81, 243, &c.
And their differences   6, 18, 54, 162, &c. are in geometrical progression.

444. Several quantities are said to be in *harmonical progression*, when, of any three which are contiguous in the series, the first is to the last, as the difference between the two first, to the difference between the two last. See art. 400.

Thus the numbers 60, 30, 20, 15, 12, 10, are in harmonical progression.

For $60 : 20 :: 60 - 30 : 30 - 20$, And $20 : 12 :: 20 - 15 : 15 - 12$,
And $30 : 15 :: 30 - 20 : 20 - 15$, And $15 : 10 :: 15 - 12 : 12 - 10$.

INFINITES and INFINITESIMALS.*

Art. 445. THE word *infinite* is used in different senses. The ambiguity of the term has been the occasion of much perplexity. It has even led to the absurd supposition, that propositions directly contradictory to each other may be mathematically demonstrated. These apparent contradictions are owing to the fact, that what is proved of infinity, when understood in one particular manner, is often thought to be true also, when the term has a very different signification. The two meanings are insensibly shifted, the one for the other, so that the proposition which is really demonstrated, is exchanged for another which is false and absurd. To prevent mistakes of this nature, it is important that the different meanings be carefully distinguished from each other.

446. INFINITE, in the highest, and perhaps the most proper sense of the word, *is that which is so great, that nothing can be added to it, or supposed to be added.*

In this sense, it is frequently used, in speaking of moral and metaphysical subjects. Thus, by infinite wisdom is meant that which will not admit of the least addition. Infinite power is that which cannot possibly be increased, even in supposition. This meaning of infinity is not applicable to the mathematics. That which is the subject of the mathematics is *quantity;* (Art. 1.) such quantity as may be conceived by the human mind. But no idea can be formed of a quantity so great that nothing can be supposed to be added to it. In this sense, an *infinite number* is inconceivable. We may increase a number by continual addition, till we obtain one that shall exceed any limits which we please to assign. By this, however, we do not arrive at a number to which

* Locke's Essays, Book 2. Chap. 17. Berkley's Analyst. Preface to Maclaurin's Fluxions. Newton's Princip. Saunderson's Algebra, Art. 336. Mansfield's Essays, Emerson's Algebra, Prob. 73.

nothing can be added; but only at one that is beyond any limits which we have hitherto set.   Farther additions may be made to it, with the same ease, as those by which it has already been increased so far.   It is therefore not infinite, in the sense in which the term has now been explained.   It is absurd to speak of the *greatest possible* number. No number can be imagined so great, as not to admit of being made greater.   We must therefore look for another meaning of infinity, before we can apply it, with propriety, to the mathematics.

447. *A mathematical quantity is said to be infinite, when it is supposed to be increased beyond any determinate limits.*

By determinate limits are meant such as can be distinctly stated.* In this sense, the natural series of numbers 1, 2, 3, 4, 5, &c. may be said to be infinite. For, if any number be mentioned ever so great, another may be supposed still greater.

The two significations of the word infinite are liable to be confounded, because they are in several points of view the same. The higher meaning includes the lower. That which is so great as to admit of no addition, must be beyond any determinate limits. But the lower does not necessarily imply the higher. Though number is capable of being increased beyond any specified limits; it will not follow, that a number can be found to which no farther additions can be made. The two infinites agree in this, that, according to each, the things spoken of are great beyond calculation. But they differ widely in another respect. To the one, nothing can be added. To the other, additions can be made at pleasure.

448. In the mathematical sense of the term, there is no absurdity in supposing *one infinite greater than another.*

We may conceive the numbers   2 2 2 2 2 2 2 &c.
            and   4 4 4 4 4 4 4 &c.

to be each extended so far as to reach round the globe, or to the most distant visible star, or beyond any greater boundary which can be mentioned. But, if the two series be equally extended, the amount of the one will be *twice as* great as the other, though both be infinite.

So, if the series $a + a^2 + a^3 + a^4 + a^5$ &c.
       and $9a + 9a^2 + 9a^3 + 9a^4 + 9a^5$ &c.

* See Note L.

be extended together beyond any specified limits, one will
be *nine* times as great as the other. But it would be absurd
to suppose one quantity greater than another, if the latter
were already so great that nothing could be added to it.

449. An infinite *number of terms* must not be mistaken for
an infinite quantity. The terms may be extended beyond
any given limits, when the amount of the whole is a finite
quantity, and even a small one. If we take half of a unit;
then half of the remainder; half of the remaining half, &c.
we shall have the series

$$\tfrac{1}{2}+\tfrac{1}{4}+\tfrac{1}{8}+\tfrac{1}{16}+\tfrac{1}{32} \ \&c.$$

in which each succeeding term is half of the preceding one.
Let the progression be continued ever so far, the sum of all
the terms can never exceed a unit. For, by the supposition,
there is still a remainder equal to the last term. And this
remainder must be added, before the amount of the whole
can be equal to a unit.

So $\tfrac{3}{2}+\tfrac{3}{4}+\tfrac{3}{8}+\tfrac{3}{16}+\tfrac{3}{32}+\tfrac{3}{64}$ &c. can never exceed 3.

450. *When a quantity is diminished till it becomes* LESS *than
any determinate quantity, it is called an* INFINITESIMAL.

Thus, in the series of fractions $\tfrac{1}{10}, \tfrac{1}{100}, \tfrac{1}{1000}, \tfrac{1}{10000}$, &c.
a unit is first divided into ten parts, then into a hundred, a
thousand, &c. One of these parts in each succeeding term,
is ten times less than in the preceding. If then the progression be continued, a portion of a unit may be obtained less
than any specified quantity. This is an infinitesimal, and, in
mathematical language, is said to be *infinitely small*. By this,
however, we are not to understand, that it can not be made
less. The same process that has reduced it below any limit
which we have yet specified, may be continued, so as to diminish it still more. And however far the progression may
be carried, we shall never arrive at a point where we must
necessarily stop.

451. In the sense now explained, mathematical quantity
may be said to be *infinitely divisible;* that is, it may be supposed to be so divided, that the *parts* shall be *less* than any
determinate quantity, and the *number* of parts *greater* than
any given number.

In the series $\tfrac{1}{10}, \tfrac{1}{100}, \tfrac{1}{1000}, \tfrac{1}{10000}$, &c. a unit is divided
into a greater and greater number of parts, till they become
infinitesimals, and the number of them infinite, that is, such a
number as exceeds any *given* number. But this does not

prove that we can ever arrive at a division in which the parts shall be the *least possible*, or the *number* of parts the *greatest possible.*

452. One infinitesimal may be *less* than another.

The series $\frac{6}{10}$, $\frac{6}{100}$, $\frac{6}{1000}$, $\frac{6}{10000}$, &c. $\Big\}$

And $\quad\quad \frac{3}{10}$, $\frac{3}{100}$, $\frac{3}{1000}$, $\frac{3}{10000}$, &c. $\Big\}$

may be carried on together, till the last term in each becomes infinitely small; and yet one of these terms will be only *half* as great as the other. For, the denominators being the same, the fractions will be as their numerators, (Art. 360. cor. 2.) that is, as 6 : 3, or 2 : 1.

Two quantities may also be divided, each into an infinite number of parts, using the term infinite in the mathematical sense, and yet the parts of one be more numerous, than those of the other.

The series $\frac{1}{10}$, $\frac{1}{100}$, $\frac{1}{1000}$, $\frac{1}{10000}$, &c. $\Big\}$

And $\quad\quad \frac{1}{40}$, $\frac{1}{400}$, $\frac{1}{4000}$, $\frac{1}{40000}$, &c. $\Big\}$

may both be infinitely extended; and yet a unit in the last series, is divided into four times as many parts as in the first. But if, by an infinite number of parts were meant such a number as could not be increased, it would be absurd to suppose the divisions of any quantity to be still more numerous.*

453. For all *practical* purposes, an infinitesimal may be considered as absolutely nothing. As it is less than any determinate quantity, it is lost even in numerical calculations. In algebraic processes, a term is often rejected as of no value, because it is infinitely small.

It is frequently expedient to admit into a calculation a small errour, or what is suspected to be an errour. It may be difficult either to avoid the objectionable part, or to ascertain its exact value, or even to determine, without a long and tedious process, whether it is really an errour or not. But if it can be shown to be infinitely small, it is of no account in practice, and may be retained or rejected at pleasure.

It is impossible to find a decimal which shall be exactly equal to the vulgar fraction $\frac{1}{3}$. Dividing the numerator by the denominator, we obtain, in the first place $\frac{3}{10}$. This is nearly equal to $\frac{1}{3}$. But $\frac{33}{100}$ is nearer, $\frac{333}{1000}$ still nearer, &c.

* See Note M.

The errour, in the first instance, is $\frac{1}{30}$.

For $\frac{3}{10} + \frac{1}{30} = \frac{9}{30} + \frac{1}{30} = \frac{10}{30} = \frac{1}{3}$.

In the same manner it may be shown, that

the difference between $\begin{cases} \frac{1}{3} \text{ and } .33, \text{ is } \frac{1}{300}, \\ \frac{1}{3} \text{ and } .333, \text{ is } \frac{1}{3000}, \&c. \end{cases}$

If the decimal be supposed to be extended beyond any assignable limit, the difference still remaining will be infinitely small. As this errour is less than any given quantity, it is of no account, and may be considered in calulation as nothing.

454. From the preceding example it will be seen, that a quantity may be continually *coming nearer* to another, and yet *never reach it.* The decimal 0.3333333 &c. by repeated additions on the right, may be made to approximate continually to $\frac{1}{3}$, but can never exactly equal it. A difference will always remain, though it may become infinitely small.

455. Though an infinitesimal is of no account *of itself*, yet its effect on other quantities is not always to be disregarded.

When it is a factor or a divisor, it may have an important influence. It is necessary, therefore, to attend to the relations which infinites, infinitesimals, and finite quantities have to each other. As an infinitesimal is less than any assignable quantity, as it is next to nothing, and, in practice, may be considered as nothing, it is frequently represented by 0.

An infinite quantity is expressed by the character ∞.

456. As an infinite quantity is incomparably greater than a finite one, the alteration of the former, by an *addition* or *subtraction* of the latter, may be disregarded in calculation. A single grain of sand is greater in comparison with the whole earth, than any finite quantity in comparison with one which is infinite. If therefore infinite and finite quantities are connected by the sign, $+$ or $-$, the latter may be rejected as of no comparative value. For the same reason, if finite quantities and infinitesimals are connected by $+$ or $-$, the latter may be expunged.

457. But if an infinite quantity be *multiplied* by one which is finite, it will be as many times increased, as any other quantity would, by the same multiplier.

If the infinite series 2 2 2 2 2 2 &c. be multiplied by 4, The product will be 8 8 8 8 8 8 &c. four times as great as the multiplicand. See art. 448.

458. And if an infinite quantity be *divided* by a finite quantity, it will be altered in the same manner as any other quantity.

If the infinite series 66666666 &c. be divided by 2,
The quotient will be 33333333 &c. half as great as the dividend.

459. If a *finite quantity* be multiplied by an *infinitesimal*, the product will be an infinitesimal; that is, putting z for a finite quantity, and 0 for an infinitesimal, (Art. 455.)

$$z \times 0 = 0.$$

If the multiplier were a *unit*, the product would be equal to the multiplicand. (Art. 90.) If the multiplier is less than a unit, the product is proportionally less. If then the multiplier is *infinitely less* than a unit, the product must be infinitely less than the multiplicand, that is, it must be an infinitesimal. Or, if an infinitesimal be considered as absolutely nothing, then the product of z into nothing is nothing. (Art. 112.)

460. On the other hand, if a finite quantity be divided by an infinitesimal, the quotient will be infinite.

$$\frac{z}{0} = \infty .$$

For, the less the divisor, the greater the quotient. If then the divisor be *infinitely* small, the quotient will be infinitely great. In other words, an infinitesimal is contained an infinite number of times in a finite quantity. This may, at first, appear paradoxical. But it is evident, that the quotient must increase, as the divisor is diminished.

Thus $6 \div 3 = 2,$ $6 \div 0.03 = 200,$
    $6 \div 0.3 = 20,$ $6 \div 0.003 = 2000,$ &c.

If then the divisor be reduced, so as to become *less* than any assignable quantity, the quotient must be *greater* than any assignable quantity.

461. If a finite quantity be divided by an infinite quantity, the quotient will be an infinitesimal.

$$\frac{z}{\infty} = 0.$$

For, the greater the divisor, the less the quotient. If

E e

then, while the dividend is finite, the divisor be infinitely great, the quotient will be infinitely small.

It must not be forgotten, that the expressions *infinitely great*, and *infinitely small* are, all along, to be understood in the *mathematical* sense, according to the definitions in arts. 447, and 450.

DIVISION by COMPOUND DIVISORS.

ART. 462. IN the section on division, the case in which the divisor is a compound quantity was omitted, because the operation, in most instances, requires some knowledge of the nature of *powers*; a subject which had not been previously explained.

Division by a compound divisor is performed by the following rule, which is substantially the same, as the rule for division in arithmetic:

To obtain the first term of the quotient, divide the first term of the dividend, by the first term of the divisor:[*]

Multiply the whole divisor, by the term placed in the quotient; subtract the product from a part of the dividend; and to the remainder bring down as many of the following terms, as shall be necessary to continue the operation:

Divide again by the first term of the divisor, and proceed as before, till all the terms of the dividend are brought down.

Ex. 1. Divide $ac+bc+ad+bd$, by $a+b$.

$a+b)ac+bc+ad+bd(c+d$
$\qquad ac+bc$, the first subtrahend.

$\qquad$ * $\qquad$ * $\qquad ad+bd$
$\qquad\qquad ad+bd$, the second subtrahend.

$\qquad$ * $\qquad$ *

Here ac, the first term of the dividend, is divided by a, the first term of the divisor, (Art. 116.) which gives c for the first term of the quotient. Multiplying the whole divisor by this, we have $ac+bc$ to be subtracted from the two first terms of the dividend. The two remaining terms are then brought

[*] See Note N.

down, and the first of them is divided by the first term of the divisor, as before. This gives d for the second term of the quotient. Then multiplying the divisor by d, we have $ad+bd$ to be subtracted, which exhausts the whole dividend, without leaving any remainder.

The rule is founded on this principle, that the product of the divisor into the several parts of the quotient, is equal to the dividend. (Art. 115.) Now by the operation, the product of the divisor into the *first* term of the quotient is subtracted from the dividend; then the product of the divisor into the *second* term of the quotient; and so on, till the product of the divisor into each term of the quotient, that is, the product of the divisor into the *whole* quotient, (Art. 100.) is taken from the dividend. If there is no remainder, it is evident that this product is *equal* to the dividend. If there is a remainder, the product of the divisor and quotient is equal to the whole of the dividend *except* the remainder. And this remainder is not included in the parts subtracted from the dividend, by operating according to the rule.

463. Before beginning to divide, it will generally be expedient to make some preparation in the *arrangement of the terms.*

The letter which is in the first term of the divisor, should be in the first term of the dividend also. And the *powers* of this letter should be arranged in order, both in the divisor and in the dividend; the highest power standing first, the next highest next, and so on.

Ex. 2. Divide $2a^2b+b^3+2ab^2+a^3$, by a^2+b^2+ab.

Here if we take a^2 for the first term of the divisor, the other terms should be arranged according to the powers of a, thus;

$$a^2+ab+b^2)a^3+2a^2b+2ab^2+b^3(a+b$$
$$a^3+\ a^2b+\ ab^2$$
$$\overline{\qquad\qquad\qquad\qquad}$$
$$a^2b+\ ab^2+b^3$$
$$a^2b+\ ab^2+b^3$$
$$\overline{\qquad\qquad\qquad\qquad}$$
$$*\qquad*\qquad*$$

In these operations, particular care will be necessary in the management of *negative quantities.* Constant attention must be paid to the rules for the signs in subtraction, multiplication and division. (Arts. 82, 105, 123.)

Ex. 3. Divide $2ax - 2a^2x - 3a^2xy + 6a^3x + axy - xy$, by $2a - y$.

If the terms be arranged according to the powers of a, they will stand thus;

$$2a - y)6a^3x - 3a^2xy - 2a^2x + axy + 2ax - xy(3a^2x - ax + x.$$
$$6a^3x - 3a^2xy$$

$$* \qquad * \qquad -2a^2x + axy$$
$$-2a^2x + axy$$

$$* \qquad * \qquad +2ax - xy$$
$$+2ax - xy$$

464. In multiplication, some of the terms, by balancing each other, may be lost in the product. (Art. 110.) These may *re-appear* in division, so as to present terms, in the course of the process, different from any which are in the dividend.

Ex. 4.

$$a + x)a^3 + x^3(a^2 - ax + x^2$$
$$a^3 + a^2x$$

$$* \quad -a^2x + x^3$$
$$-a^2x - ax^2$$

$$* \qquad ax^2 + x^3$$
$$ax^2 + x^3$$

Ex. 5.

$$a^2 - 2ax + 2x^2)a^4 + 4x^4(a^2 + 2ax + 2x^2$$
$$a^4 - 2a^3x + 2a^2x^2$$

$$* \quad +2a^3x - 2a^2x^2 + 4x^4$$
$$+2a^3x - 4a^2x^2 + 4ax^3$$

$$* \qquad +2a^2x^2 - 4ax^3 + 4x^4$$
$$+2a^2x^2 - 4ax^3 + 4x^4$$

If the learner will take the trouble to multiply the quotient into the divisor, in the two last examples, he will find, in the partial products, the several terms which appear in the process of dividing. But most of them, by balancing each other, are lost in the general product.

Ex. 6.

$$a+1)a^3+a^2+a^2b+ab+3ac+3c(a^2+ab+3c$$
$$a^3+a^2$$

$$* \quad * \quad a^2b+ab$$
$$a^2b+ab$$

$$* \quad * \quad 3ac+3c$$
$$3ac+3c$$

Ex. 7.

$$a+b-c(a+b-c-ax-bx+cx(1-x$$
$$a+b-c$$

$$* \; * \; *-ax-bx+cx$$
$$-ax-bx+cx$$

Ex. 8. Divide $2a^4-13a^3x+11a^2x^2-8ax^3+2x^4$, by $2a^2-ax+x^2$.   Quotient. $a^2-6ax+2x^2$.

465. When there is a *remainder* after all the terms of the dividend have been brought down, this may be placed over the divisor and added to the quotient, as in arithmetic.

Ex. 9.

$$a+b)ac+bc+ad+bd+x(c+d+\frac{x}{a+b}.$$
$$ac+bc$$

$$* \quad * \quad ad+bd$$
$$ad+bd$$

$$* \quad * \quad x$$

Ex. 10.

$$d-h)ad-ah+bd-bh+y(a+b+\frac{y}{d-h}\cdot$$
$$ad-ah$$

$$*\quad*\quad bd-bh$$
$$bd-bh$$

$$*\quad*\quad y$$

It is evident that $a+b$ is the quotient belonging to the whole of the dividend, *excepting* the remainder y. (Art. 562.) And $\frac{y}{d-h}$ is the quotient belonging to this remainder. (Art. 124.)

Ex. 11. Divide $6ax+2xy-3ab-by+3ac+cy+h$, by $3a+y$. Quotient. $2x-b+c+\dfrac{h}{3a+y}\cdot$

Ex. 12. Divide $a^2b-3a^2+2ab-6a-4b+22$, by $b-3$.
Quotient. $a^2+2a-4+\dfrac{10}{b-3}\cdot$

Ex. 13. See art. 283.

$$a+\sqrt{b})ac+c\sqrt{b}+a\sqrt{d}+\sqrt{bd}(c+\sqrt{d}.$$
$$ac+c\sqrt{b}$$

$$*\quad*\quad a\sqrt{d}+\sqrt{bd}$$
$$a\sqrt{d}+\sqrt{bd}$$

Ex. 14.

$$a+\sqrt{y})a+\sqrt{y}+ar\sqrt{y}+ry(1+r\sqrt{y}$$
$$a+\sqrt{y}$$

$$*\quad*\quad ar\sqrt{y}+ry$$
$$ar\sqrt{y}+ry$$

466. A regular series of quotients is obtained, by dividing the difference of the *powers* of two quantities, by the difference of the quantities. Thus,

$$(y^2 - a^2) \div (y - a) = y + a,$$
$$(y^3 - a^3) \div (y - a) = y^2 + ay + a^2,$$
$$(y^4 - a^4) \div (y - a) = y^3 + ay^2 + a^2 y + a^3,$$
$$(y^5 - a^5) \div (y - a) = y^4 + ay^3 + a^2 y^2 + a^3 y + a^4,$$
&c.

Here it will be seen, that the index of y, in the first term of the quotient, is less by 1, than in the dividend; and that it decreases by 1, from the first term to the last but one:

While the index of a increases by 1, from the second term to the last, where it is less by 1, than in the dividend.

This may be expressed in a general formula, thus,

$$(y^m - a^m (\div (y - a) = y^{m-1} + ay^{m-2} \ldots + a^{m-2} y + a^{m-1}.$$

To demonstrate this, we have only to multiply the quotient into the divisor. (Art. 115.)

All the terms except two, in the partial products, will be balanced by each other; and will leave the general product the same as the dividend.

Mult. $y^4 + ay^3 + a^2 y^2 + a^3 y + a^4$
Into $y \ - a$

$$y^5 + ay^4 + a^2 y^3 + a^3 y^2 + a^4 y$$
$$- ay^4 - a^2 y^3 - a^3 y^2 - a^4 y - a^5$$

Product. y^5 *   *   *   * $- a^5$.

So mult. $y^{m-1} + ay^{m-2} + a^2 y^{m-3} \ldots + a^{m-2} y + a^{m-1}$
Into $y - a$

$$y^m + ay^{m-1} + a^2 y^{m-2} \ldots + a^{m-2} y^2 + a^{m-1} y$$
$$- ay^{m-1} - a^2 y^{m-2} \ldots - a^{m-2} y^2 - a^{m-1} y - a^m$$

Prod. y^m *   *   *   * $- a^m$.

INVOLUTION and EXPANSION of BINOMIALS.[*]

ART. 467. THE manner in which a binomial, as well as any other compound quantity, may be involved by repeated multiplications, has been shown in the section on powers. (Art. 213.) But when a high power is required, the operation becomes long and tedious.

This has led mathematicians to seek for some general principle, by which the involution may be more easily and expeditiously performed. We are chiefly indebted to Sir Isaac Newton for the method which is now in common use. It is founded on what is called the *Binomial Theorem*, the invention of which was deemed of such importance to mathematical investigation, that it is engraved on his monument in Westminster Abbey.

468. If the binomial root be $a+b$, we may obtain, by multiplication, the following powers. (Art. 213.)

$$(a+b)^2 = a^2 + 2ab + b^2$$
$$(a+b)^3 = a^3 + 3a^2b + 3ab^2 + b^3$$
$$(a+b)^4 = a^4 + 4a^3b + 6a^2b^2 + 4ab^3 + b^4$$
$$(a+b)^5 = a^5 + 5a^4b + 10a^3b^2 + 10a^2b^3 + 5ab^4 + b^5, \&c.$$

By attending to this series of powers, we shall find, that the *exponents* preserve an invariable order through the whole. This will be very obvious, if we take the exponents by themselves, unconnected with the letters to which they belong.

[*] Simpson's Algebra, Sec. 15. Simpson's Fluxions, Art. 99. Euler's Algebra, Sec. 2. Chap. 10. Manning's Algebra. Saunderson's Algebra, Art. 380. Vince's Fluxions, Art. 33. Waring's Med. Anal. p. 415. Lacroix's Algebra, Art. 135. Do..Comp. Art. 70. Lond. Phil. Trans. 1795.

F f

In the square, the exponents $\begin{cases} \text{of } a \text{ are } 2, 1, 0 \\ \text{of } b \text{ are } 0, 1, 2 \end{cases}$

In the cube, the exponents $\begin{cases} \text{of } a \text{ are } 3,2,1,0 \\ \text{of } b \text{ are } 0,1,2,3 \end{cases}$

In the 4th power, the exponents $\begin{cases} \text{of } a \text{ are } 4,3,2,1,0 \\ \text{of } b \text{ are } 0,1,2,3,4 \end{cases}$

In the 5th power, the exponents $\begin{cases} \text{of } a \text{ are } 5,4,3,2,1,0 \\ \text{of } b \text{ are } 0,1,2,3,4,5, \text{ &c.} \end{cases}$

Here it will be seen, at once, that the exponents, of a in the *first* term, and of b in the *last*, are each equal to the index of the power; and that the *sum* of the exponents of the two letters is in every term the same. Thus in the fifth power,

The sum of the exponents $\begin{cases} \text{in the first term, is } 5+0=5 \\ \text{in the second,} \quad 4+1=5 \\ \text{in the third,} \quad 3+2=5, \text{&c:} \end{cases}$

It is farther to be observed, that the exponents of a regularly *decrease* to 0, and that the exponents of b *increase* from 0. That this will universally be the case, to whatever extent the involution may be carried, will be evident, if we consider, that, in raising from any power to the next, *each term* is multiplied both by a and by b.

Thus $(a+b)^2 = a^2 + 2ab + b^2$
Mult. by $\qquad a+b$

$\overline{\qquad\qquad\qquad\qquad}$ [of a in each term.
$a^3 + 2a^2b + ab^2$, Here 1 is added to the exp.
$\qquad a^2b + 2ab^2 + b^3$, Here 1 is added to the
$\overline{\qquad\qquad\qquad\qquad}$ [exp. of b in each term.
$(a+b)^3 = a^3 + 3a^2b + 3ab^2 + b^3$

If the exponents, before the multiplication, increase and decrease by 1, and if the multiplication adds 1 to each, it is evident they must still increase and decrease in the same manner as before.

469. If then $a+b$ be raised to a power whose exponent is n,

The exp's of a will be $\quad n, n-1, n-2, \ldots \ 2, \quad 1, \quad 0;$
And the exp's of b will be $0, \quad 1, \quad 2, \quad \ldots n-2, n-1, n.$

The terms in which a power is expressed, consist of the letters with their *exponents*, and the *co-efficients*. Setting aside the co-efficients for the present, we can determine, from the preceding observations, the letters and exponents of any power whatever.

Thus the 8th power of $a+b$, when written without the co-efficients, is

$$a^8 + a^7 b + a^6 b^2 + a^5 b^3 + a^4 b^4 + a^3 b^5 + a^2 b^6 + ab^7 + b^8.$$

And the nth power of $a+b$ is,

$$a^n + a^{n-1}b + a^{n-2}b^2 \ldots a^2 b^{n-2} + ab^{n-1} + b^n.$$

470. The *number* of terms is greater by 1, than the index of the power. For, if the index of the power is n, a has, in different terms, every index from n down to 1; and there is one additional term which contains only b. Thus,

The square has 3 terms,   The 4th power, 5,
The cube   4,   The 5th power, 6, &c.

471. The next step is to find the *co-efficients*. This part of the subject is more complicated.

In the series of powers at the beginning of art. 468, the co-efficients, taken separate from the letters, are as follows;

In the square,   1, 2, 1,   whose sum is $4 = 2^2$,
In the cube,   1, 3, 3, 1, $8 = 2^3$,
In the 4th power,   1, 4, 6, 4, 1, $16 = 2^4$,
In the 5th power, 1, 5, 10, 10, 5, 1, $32 = 2^5$.

The order which these co-efficients observe is not obvious, like that of the exponents, upon a bare inspection. But they will be found on examination to be all subject to the following law;

472. The co-efficient of the first term is 1; that of the second is equal to the index of the power; and universally, if the co-efficient of any term be multiplied by the index of the leading quantity in that term, and divided by the index of the following quantity increased by 1, it will give the co-efficient of the succeeding term.*

Of the two letters in a term, the first is called the *leading* quantity, and the other, the *following* quantity. In the examples which have been given in this section, a is the leading quantity, and b the following quantity.

It may frequently be convenient to represent the co-efficients, in the several terms, by the capital letters, $A, B, C,$ &c.

* See Note ·O.

The nth power of $a+b$, without the co-efficients, is

$$a^n + a^{n-1}b + a^{n-2}b^2 + a^{n-3}b^3 + a^{n-4}b^4, \&c. \text{ (Art. 469.)}$$

And the co-efficients are,

$A = n$, the co-efficient of the *second* term;

$$B = n \times \frac{n-1}{2}, \qquad \text{of the } third \text{ term;}$$

$$C = n \times \frac{n-1}{2} \times \frac{n-2}{3}, \text{ of the } fourth \text{ term;}$$

$$D = n \times \frac{n-1}{2} \times \frac{n-2}{3} \times \frac{n-3}{4}, \text{ of the } fifth \text{ term, \&c.}$$

The regular manner in which these co-efficients are derived one from another, will be readily perceived.

473. By recurring to the numbers in art. 471, it will be seen, that the co-efficients first *increase*, and then *decrease* at the same rate; so that they are equal, in the first term and the last, in the second and last but one, in the third and last but two; and, universally, in any two terms equally distant from the extremes. The reason of this is, that $(a+b)^n$ is the same as $(b+a)^n$; and if the order of the terms in the binomial root be changed, the whole series of terms in the power will be inverted.

It is sufficient, then, to find the co-efficients of *half* the terms. These repeated, will serve for the whole.

474. In any power of $(a+b)$, the *sum* of the co-efficients is equal to the number 2 raised to that power. See the list of co-efficients in art. 471. The reason of this is, that, according to the rules of multiplication, when any quantity is involved, the *letters* are multiplied into each other, and the *co-efficients* into each other. Now the co-efficients of $a+b$ being $1+1=2$, if these be involved, a series of the powers of 2 will be produced.

Multiplying $1+1$	or 2
Into $1+1$	2

The square is $1+2+1$	or $4=$ the square of 2
Mult. again $1+1$	2

The cube is $1+3+3+1$	or $8=$ the cube of 2, &c.

475. The principles which have now been explained may mostly be comprised in the following general theorem, called

The Binomial Theorem.

The index of the leading quantity of the power of a binomial, begins in the first term with the index of the power, and decreases regularly by 1. The index of the following quantity begins with 1 in the second term, and increases regularly by 1. (Art. 468.)

The co-efficient of the first term is 1; that of the second is equal to the index of the power; and universally, if the co-efficient of any term be multiplied by the index of the leading quantity in that term, and divided by the index of the following quantity increased by 1, it will give the co-efficient of the succeeding term. (Art. 472.)

In algebraic characters, the theorem is

$$(a+b)^n = a^n + n \times a^{n-1}b + n \times \frac{n-1}{2}a^{n-2}b^2, \&c.$$

It is here supposed, that the *terms* of the binomial have no other co-efficients or exponents than 1. Other binomials may be reduced to this form by substitution.

Ex. 1. What is the 6th power of $x+y$?
The terms without the co-efficients, are

$$x^6, x^5y, x^4y^2, x^3y^3, x^2y^4, xy^5, y^6.$$

And the co-efficients are

$$1, \quad 6, \quad \frac{6\times5}{2}, \quad \frac{15\times4}{3}, \quad \frac{20\times3}{4}, \quad 6, \quad 1$$

that is, $\cdot$1, 6, 15, 20, 15, 6, 1.

Prefixing these to the several terms, we have the power required;

$$x^6 + 6x^5y + 15x^4y^2 + 20x^3y^3 + 15x^2y^4 + 6xy^5 + y^6.$$

2. $(d+h)^5 = d^5 + 5d^4h + 10d^3h^2 + 10d^2h^3 + 5dh^4 + h^5.$

3. What is the nth power of $b+y$?

Ans. $b^n + Ab^{n-1}y + Bb^{n-2}y^2 + Cb^{n-3}y^3 + Db^{n-4}y^4$, &c.

That is, supplying the co-efficients which are here represented by A, B, C, &c. (Art. 472.)

$$b^n + n \times b^{n-1}y + n \times \frac{n-1}{2} \times b^{n-2}y^2, \&c.$$

476. A *residual* quantity may be involved in the same manner, without any variation, except in the *signs*. By repeated multiplications, as in art. 213, we obtain the following powers of $(a-b)$.

$$(a-b)^2 = a^2 - 2ab + b^2$$
$$(a-b)^3 = a^3 - 3a^2b + 3ab^2 - b^3.$$
$$(a-b)^4 = a^4 - 4a^3b + 6a^2b^2 - 4ab^3 + b^4, \&c.$$

By comparing these with the like powers of $(a+b)$ in art. 468, it will be seen, that there is no difference, except in the *signs*. There, *all* the terms are positive. Here, the terms which contain the *odd* powers of b are negative. See art. 218.

The sixth power of $(x-y)$ is

$$x^6 - 6x^5y + 15x^4y^2 - 20x^3y^3 + 15x^2y^4 - 6xy^5 + y^6.$$

The nth power of $(a-b)$ is

$$a^n - Aa^{n-1}b + Ba^{n-2}b^2 - Ca^{n-3}b^3, \&c.$$

477. When one of the terms of a binomial is *a unit*, it is generally omitted in the power, except in. the first or last term; because every power of 1 is 1, (Art. 209.) and this, when it is a factor, has no effect upon the quantity with which it is connected. (Art. 90.)

Thus the cube of $(x+1)$ is $x^3 + 3x^2 \times 1 + 3x \times 1^2 + 1^3$,
Which is the same as $x^3 + 3x^2 + 3x + 1$.

The insertion of the powers of 1 is of no use, unless it be to preserve the *exponents* of both the leading and the following quantity in each term, for the purpose of finding the co-efficients. But this will be unnecessary, if we bear in mind, that the *sum* of the two exponents, in each term, is equal to the index of the power. (Art. 468.) So that, if we have the exponent of the *leading* quantity, we may know that of the *following* quantity, and v. v.

Ex. 1. The sixth power of $(1-y)$ is
$$1-6y+15y^2-20y^3+15y^4-6y^5+y^6.$$

2. $(1+x)^n = 1+Ax^2+Bx^3+Cx^4+Dx^5$, &c.

478. From the comparatively simple manner in which the power is expressed, when the first term of the root is a unit, is suggested the expediency of reducing other binomials to this form.

The quotient of $(a+x)$ divided by a is $\left((1+\dfrac{x}{a}\right)$. This multiplied into the divisor, is equal to the dividend; that is,

$$(a+x)=a\times\left(1+\frac{x}{a}\right) \text{ therefore } (a+x)^n=a^n\times\left(1+\frac{x}{a}\right)^n.$$

By expanding the factor $\left(1+\dfrac{x}{a}\right)^n$, we have

$$(a+x)^n=a^n\times\left(1+\frac{x}{a}\right)^n=a^n\times\left(1+A\frac{x}{a}+B\frac{x^2}{a^2}+C\frac{x^3}{a^3}, \&c.\right)$$

479. When the index of the power to which any binomial is to be raised is a *positive whole number,* the series will *terminate.* The number of terms will be limited, as in all the preceding examples.

For, as the index of the leading quantity continually decreases by 1, it must, in the end, become 0, and then the series will break off.

Thus, the 5 term of the fourth power of $a+x$ is x^4, or a^0x^4, a^0 being commonly omitted, because it is equal to 1. (Art. 207.)   If we attempt to continue the series farther, the co-efficient of the next term, according to the rule, will be $\dfrac{1\times0}{5}=0$. (Art. 112.)   And as the co-efficients of all succeeding terms must depend on this, they will also be 0.

480. If the index of the proposed power is *negative,* this can never become 0, by the successive subtractions of a unit. The series will, therefore, *never terminate;* but, like many decimal fractions, may be continued to any extent that is desired.

Ex. Expand into a series $\dfrac{1}{(a+y)^2} = (a+y)^{-2}$.

The terms, without the co-efficients, are

$$a^{-2},\ a^{-3}y,\ a^{-4}y^2,\ a^{-5}y^3,\ a^{-6}y^4,\ \&c.$$

The co-ef. of the 2d term is -2, of the 4th $\dfrac{+3 \times -4}{3} = -4$,

of the 3d, $\dfrac{-2 \times -3}{2} = +3$, of the 5th $\dfrac{-4 \times -5}{4} = +5$.

The series then is

$$a^{-2} - 2a^{-3}y + 3a^{-4}y^2 - 4a^{-5}y^3 + 5a^{-6}y^4,\ \&c.$$

Here the law of the progression is apparent; the co-efficients increase regularly by 1, and their signs are alternately positive and negative.

481. The Binomial Theorem is of great utility, not only in raising powers, but particularly in finding the *roots* of binomials. A root may be expressed in the same manner as a power, except that the exponent is, in the one case an *integer*, in the other a fraction. (Art. 245.) Thus $(a+b)^n$ may be either a power or a root. It is a power if $n=2$, but a root if $n=\frac{1}{2}$.

482. If a root be expanded by the binomial theorem, the series *will never terminate*. A series produced in this way terminates, only when the index of the leading quantity becomes equal to 0, so as to destroy the co-efficients of the succeeding terms. (Art. 479.) But, according to the theorem, the difference in the index, between one term and the next, is always a unit; and a *fraction*, though it may change from positive to negative, can not become exactly equal to 0, by successive subtractions of a unit. Thus, if the index in the first term be $\frac{1}{2}$, it will be,

In the 2d, $\frac{1}{2} - 1 = -\frac{1}{2}$,   In the 4th, $-\frac{3}{2} - 1 = -\frac{5}{2}$,

In the 3d, $-\frac{1}{2} - 1 = -\frac{3}{2}$,   In the 5th, $-\frac{5}{2} - 1 = -\frac{7}{2}$, &c.

Ex. What is the square root of $(a+b)$?

The terms, without the co-efficients, are

$$a^{\frac{1}{2}},\ a^{-\frac{1}{2}}b,\ a^{-\frac{3}{2}}b^2,\ a^{-\frac{5}{2}}b^3,\ a^{-\frac{7}{2}}b^4,\ \&c.$$

The co-efficient of the 2d term is $+\frac{1}{2}$,

of the 3d, $\dfrac{\frac{1}{2} \times -\frac{1}{2}}{2} = -\frac{1}{8}$,   of the 4th, $\dfrac{-\frac{1}{2} \times -\frac{3}{2}}{3} = +\frac{1}{16}$

And the series is $a^{\frac{1}{2}} + \frac{1}{2}a^{-\frac{1}{2}}b - \frac{1}{8}a^{-\frac{3}{2}}b^2 + \frac{1}{16}a^{-\frac{5}{2}}b^3 \&c.$

483. The binomial theorem may also be applied to quantities consisting of *more than two terms*. By substitution, several terms may be reduced to two, and when the compound expressions are restored, such of them as have exponents may be separately expanded.

Ex. What is the cube of $a+b+c$?

Substituting h for $(b+c)$, we have $a+(b+c)=a+h$.

And by the theorem, $(a+h)^3=a^3+3a^2h+3ah^2+h^3$.

That is, restoring the value of h,

$$(a+b+c)^3=a^3+3a^2\times(b+c)+3a\times(b+c)^2+(b+c)^3.$$

The two last terms contain powers of $(b+c)$; but these may be separately involved.

EVOLUTION of COMPOUND QUANTITIES.

ART. 484. THE roots of compound quantities may be extracted by the following general rule:

After arranging the terms according to the powers of one of the letters, so that the highest power shall stand first, the next highest next, &c.

Take the root of the first term, for the first term of the required root:

Subtract the power from the given quantity, and divide the first term of the remainder, by the first term of the root involved to the next inferiour power, and multiplied by the index of the given power;†the quotient will be the next term of the root.

Subtract the power of the terms already found from the given quantity, and, using the same divisor, proceed as before.

This rule verifies itself. For the root, whenever a new term is added to it, is involved, for the purpose of subtracting its power from the given quantity; and when the power is *equal* to this quantity, it is evident the true root is found.

Ex. 1. Extract the cube root of

$a^6 + 3a^5 - 3a^4 - 11a^3 + 6a^2 + 12a - 8 (a^2 + a - 2$
a^6, the first subtrahend.

$3a^4)$ * $3a^5$, &c. the first remainder.

$a^6 + 3a^5 + 3a^4 + a^3$, the 2d subtrahend.

$3a^4)$ *   * $-6a^4$, &c. the 2d remainder.

$a^6 + 3a^5 - 3a^4 - 11a^3 + 6a^2 + 12a - 8.$

---

† By the *given power* is meant a power of the same name with the required root. As powers and roots are correlative, any quantity is the square of its square root, the cube of its cube root, &c.

Here a^2, the cube root of a^6, is taken for the first term of the required root. The power a^6 is subtracted from the given quantity. For a divisor, the first term of the root is squared, that is, raised to the next inferiour power, and multiplied by 3, the index of the given power.

By this, the first term of the remainder $3a^5$, &c. is divided; and the quotient a is added to the root. Then $a^2 + a$, the part of the root now found, is involved to the cube; for the second subtrahend, which is subtracted from the whole of the given quantity. The first term of the remainder $-6a^4$, &c. is divided by the divisor used above, and the quotient -2 is added to the root. Lastly, the whole root is involved to the cube, and the power is found to be exactly equal to the given quantity.

It is not necessary to write the remainders at length, as, in dividing, the first term only is wanted.

2. Extract the fourth root of
$$a^4 + 8a^3 + 24a^2 + 32a + 16(a + 2$$
$$a^4$$

$$4a^3 (* \quad 8a^3, \&c.$$

$$a^4 + 8a^3 + 24a^2 + 32a + 16.$$

3. What is the 5th root of
$$a^5 + 5a^4 b + 10a^3 b^2 + 10a^2 b^3 + 5ab^4 + b^5 ?$$

Ans. $a + b$.

4. What is the cube root of
$$a^3 - 6a^2 b + 12ab^2 - 8b^3 ?$$
Ans. $a - 2b$.

5. What is the square root of
$$4a^2 - 12ab + 9b^2 + 16ah - 24bh + 16h^2 (2a - 3b + 4h$$
$$4a^2$$

$$4a) * \quad -12ab, \&c.$$

$$4a^2 - 12ab + 9b^2$$

$$4a) * \quad * \quad * + 16ah, \&c.$$

$$4a^2 - 12ab + 9b^2 + 16ah - 24bh + 16h^2.$$

In finding the divisor here, the term $2a$ in the root is not involved, because the power next below the square is the first power.

485. But the square root is more commonly extracted by the following rule, which is of the same nature, as that which is used in arithmetic.

After arranging the terms according to the powers of one of the letters, take the root of the first term, for the first term of the required root, and subtract the power from the given quantity.

Bring down two other terms for a dividend. Divide by double the root already found, and add the quotient, both to the root, and to the divisor. Multiply the divisor thus increased, into the term last placed in the root, and subtract the product from the dividend.

Bring down two or three additional terms, and proceed as before.

Ex. 1. What is the square root of
$$a^2+2ab+b^2+2ac+2bc+c^2\,(a+b+c$$
a^2, the first subtrahend.

$2a+b)$ * $2ab+b^2$
Into $b =$ $2ab+b^2$, the 2d subtrahend.

$2a+2b+c)$ *   * $2ac+2bc+c^2$
Into $c =$ $2ac+2bc+c^2$, the 3d subtrahend.

Here it will be seen, that the several subtrahends are successively taken from the given quantity, till it is exhausted. If then, these subtrahends are together equal to the square of the terms placed in the root, the root is truly assigned by the rule.

The *first* subtrahend is the square of the first term of the root.

The *second* subtrahend is the product of the second term of the root, into itself, and into twice the preceding term.

The *third* subtrahend is the product of the third term of the root, into itself, and into twice the sum of the two preceding terms, &c.

That is, the subtrahends are equal to
$$a^2+(2a+b)\times b+(2a+2b+c)\times c,\ \&c.$$
and this expression is equal to the square of the root.

For $(a+b)^2 = a^2 + 2ab + b^2 = a^2 + (2a+b) \times b$. (Art.120.)

And putting $h = a+b$, the square $h^2 = a^2 + (2a+b) \times b$.

And $(a+b+c)^2 = (h+c)^2 = h^2 + (2h+c) \times c$;

that is, restoring the values of h and h^2,

$$(a+b+c)^2 = a^2 + (2a+b) \times b + (2a+2b+c) \times c.$$

In the same manner it may be proved, that, if another term be added to the root, the power will be increased, by the product of that term, into itself, and into twice the sum of the preceding terms.

The demonstration will be substantially the same, if some of the terms be *negative*.

2. What is the square root of

$$1 - 4b + 4b^2 + 2y - 4by + y^2 \quad (1 - 2b + y$$

$$1$$

$$\overline{}$$

$$2 - 2b) \,* - 4b + 4b^2$$
$$\text{Into} -2b = -4b + 4b^2$$

$$\overline{}$$

$$2 - 4b + y) \,* \qquad * \quad 2y - 4by + y^2$$
$$\text{Into} \qquad y = \qquad 2y - 4by + y^2.$$

$$\overline{}$$

3. What is the square root of

$$a^6 - 2a^5 + 3a^4 - 2a^3 + a^2 ? \qquad \text{Ans. } a^3 - a^3 + a.$$

4. What is the square root of

$$a^4 + 4a^2 b + 4b^2 - 4a^2 - 8b + 4 ? \qquad \text{Ans. } a^2 + 2b - 2.$$

486. It will frequently facilitate the extraction of roots, to consider the index as composed of two or more *factors*.

Thus $a^{\frac{1}{4}} = a^{\frac{1}{2} \times \frac{1}{2}}$. (Art. 258.) And $a^{\frac{1}{6}} = a^{\frac{1}{3} \times \frac{1}{2}}$. That is,

The fourth root is equal to the square root of the square root;

The sixth root is equal to the square root of the cube root;

The eighth root is equal to the square root of the fourth root, &c.

To find the sixth root, therefore, we may first extract the cube root, and then the square root of this.

INFINITE SERIES.

ART. 487.　IT is frequently the case, that, in attempting to extract the root of a quantity, or to divide one quantity by another, we find it impossible to assign the quotient or root with exactness. But, by continuing the operation, one term after another may be added, so as to bring the result nearer and nearer to the value required. See art. 454. When the number of terms is supposed to be extended beyond any determinate limits, the expression is called an *infinite series*. The *quantity*, however, may be finite, though the number of terms be unlimited.

An infinite series may appear, at first view, much less simple, than the expression from which it is derived. But the former is, frequently, more within the power of calculation than the latter. Much of the labour and ingenuity of mathematicians has, accordingly, been employed on the subject of series. If it were necessary to find each of the terms by actual calculation, the undertaking would be hopeless. But a few of the leading terms will, generally, be sufficient to determine the law of the progression.

488. A fraction may often be expanded into an infinite series, *by dividing the numerator by the denominator.* For the *value* of a fraction is equal to the quotient of the numerator divided by the denominator. (Art. 135.) When this quotient can not be expressed in a limited number of terms, it may be represented by an infinite series.

Ex. 1. To reduce the fraction $\frac{1}{1-a}$ to an infinite series, divide 1 by $1-a$, according to the rule in art. 462.

$$1-a)1 \quad (1+a+a^2+a^3 \ \&c.$$
$$\underline{1-a}$$

$$* \quad a$$
$$a-a^2$$

$$* \quad a^2$$
$$a^2-a^3$$

$$* \quad a^3 \ \&c.$$

By continuing the operation, we obtain the terms
$1+a+a^2+a^3+a^4+a^5+a^6$, &c. which are sufficient
to show that the series, after the first term, consists of the
powers of a, rising regularly one above another.

That the series may *converge*, that is, come nearer
and nearer to the exact value of the fraction, it is necessary that the first term of the divisor be greater than the
second. In the example just given, 1 must be greater than
a. For, at each step of the division, there is a *remainder;*
and the quotient is not complete, till this is placed over the
divisor and annexed. Now the first remainder is a, the second a^2, the third a^3, &c. If a then is greater than 1, the
remainder continually increases; which shows, that, the farther the division is carried, the greater is the quantity, either
positive or negative, which ought to be added to the quotient. The series is, therefore, *diverging* instead of *con-
verging.*

But if a be *less* than 1, the remainders a, a^2, a^3, &c. will
continually decrease. For powers are raised by multiplication; and if the multiplier be less than a unit, the product
will be less than the multiplicand. (Art. 90.) If a be taken
equal to $\frac{1}{2}$, then by art. 223,

$$a^2=\tfrac{1}{4}, \qquad a^3=\tfrac{1}{8}, \qquad a^4=\tfrac{1}{16}, \qquad a^5=\tfrac{1}{32}, \ \&c.$$

and we have

$$\frac{1}{1-\frac{1}{2}}=\frac{1}{\frac{1}{2}}=2=1+\tfrac{1}{2}+\tfrac{1}{4}+\tfrac{1}{8}+\tfrac{1}{16}+\tfrac{1}{32}+\tfrac{1}{64} \ \&c.$$

Here, the *two* first terms $=1+\frac{1}{2}$, which is less than 2, by $\frac{1}{2}$;
the *three* first $\quad =1+\frac{3}{4}$, $\qquad$ less than 2, by $\frac{1}{4}$;
the *four* first $\quad =1+\frac{7}{8}$, $\qquad$ less than 2, by $\frac{1}{8}$;
$\qquad\qquad\qquad\qquad\qquad\qquad\qquad\qquad$ [&c.

So that, the farther the series is carried, the nearer it approaches to the value of the given fraction, which is equal to 2.

2. If $\dfrac{1}{1+a}$ be expanded, the series will be the same as that from $\dfrac{1}{1-a}$, except that the terms which consist of the *odd* powers of a will be *negative*.

So that $\dfrac{1}{1+a} = 1 - a + a^2 - a^3 + a^4 - a^5 + a^6$ &c.

3. Reduce $\dfrac{h}{a-b}$ to an infinite series.

$$a-b \overline{)h} \quad \left(\frac{h}{a} + \frac{bh}{a^2} + \frac{b^2h}{a^3} \right. \text{ &c.}$$
$$h - \frac{bh}{a}$$
$$\ast \quad \frac{bh}{a}$$
$$\frac{bh}{a} - \frac{b^2h}{a^2}$$
$$\frac{b^2h}{a^2} \text{ &c.}$$

Here h divided by a, gives $\dfrac{h}{a}$ for the first term of the quotient. (Art. 124.) This is multiplied into $a-b$, and the product is $h - \dfrac{bh}{a}$; (Arts. 159, 158.) which subtracted from h leaves $\dfrac{bh}{a}$. This divided by a, gives $\dfrac{bh}{a^2}$ (Art. 163.) for the second term of the quotient. If the operation be continued in the same manner, we shall obtain the series,

$$\frac{h}{a} + \frac{bh}{a^2} + \frac{b^2h}{a^3} + \frac{b^3h}{a^4} + \frac{b^4h}{a^5} \text{ &c.}$$

in which the exponents of b and of a increase regularly by 1.

4. Reduce $\dfrac{1+a}{1-a}$ to an infinite series.

Ans. $1+2a+2a^2+2a^3+2a^4$ &c.

489. An infinite series may be produced, *by extracting the root of a compound surd.*

Ex. I. Reduce $\sqrt{a^2+b^2}$ to an infinite series, by extracting the square root, according to the rule in art. 485.

$$a^2+b^2 \left(a+\frac{b^2}{2a}-\frac{b^4}{8a^3}+\frac{b^6}{16a^5} \text{ &c.} \right.$$

$$a^2$$

$$\overline{}$$

$$2a+\frac{b^2}{2a} \Big) \quad * \quad b^2$$

$$b^2+\frac{b^4}{4a^2}$$

$$\overline{\phantom{b^2+\frac{b^4}{4a^2}}}$$

$$2a+\frac{b^2}{a}-\frac{b^4}{8a^3} \Big) -\frac{b^4}{4a^2} \text{ &c.}$$

Here a, the root of the first term, is taken for the first term of the series; and the power a^2 is subtracted from the given quantity. The remainder b^2 is divided by $2a$, which gives $\dfrac{b^2}{2a}$, for the second term of the root. (Art. 124.) The divisor, with this term added to it, is then multiplied into the term, and the product is $b^2+\dfrac{b^4}{4a^2}$. (Arts. 159, 155.) This subtracted from b^2 leaves $-\dfrac{b^4}{4a^2}$, which divided by $2a$ gives $-\dfrac{b^4}{8a^3}$, for the third term of the root. (Art. 163.) &c.

2. $\sqrt{a^2-b^2}=a-\dfrac{b^2}{2a}-\dfrac{b^4}{8a^3}-\dfrac{b^6}{16a^5}$ &c.

3. $\sqrt{2}=\sqrt{1+1}=1+\frac{1}{2}-\frac{1}{8}+\frac{1}{16}$ &c.

H h

490. A binomial which has a negative or fractional exponent, may be expanded into an infinite series by the *binomial theorem*. See arts. 480, 482.

Ex. 1. Expand $\dfrac{1}{(a+y)^4} = (a+y)^{-4}$ into an infinite series.

The terms, without the co-efficients, are

$$a^{-4}, \ a^{-5}y, \ a^{-6}y^2, \ a^{-7}y^3, \ a^{-8}y^4 \ \&c.$$

The co-ef. of the 2d term, is −4, of the 4th, $\dfrac{+10 \times -6}{3} = -20$,

of the 3d, $\dfrac{-4 \times -5}{2} = +10$, of the 5th, $\dfrac{-20 \times -7}{4} = +35$.

The series then is

$$a^{-4} - 4a^{-5}y + 10a^{-6}y^2 - 20a^{-7}y^3 + 35a^{-8}y^4 \ \&c.$$

Which is the same (Art. 207.) as

$$\frac{1}{a^4} - \frac{4y}{a^5} + \frac{10y^2}{a^6} - \frac{20y^3}{a^7} + \frac{35y^4}{a^8} \ \&c.$$

2. Expand $(x+y)^{\frac{1}{2}}$ into an infinite series. See art. 482.

Ans. $x^{\frac{1}{2}} + \frac{1}{2}x^{-\frac{1}{2}}y - \frac{1}{8}x^{-\frac{3}{2}}y^2 + \frac{1}{16}x^{-\frac{5}{2}}y^3 \ \&c.$

Which is the same as

$$x^{\frac{1}{2}} + \frac{y}{2x^{\frac{1}{2}}} - \frac{y^2}{8x^{\frac{3}{2}}} + \frac{y^3}{16x^{\frac{5}{2}}} \ \&c.$$

491. Though an infinite series consists of an unlimited number of terms, yet, in many cases, it is not difficult to find what is called the *sum of the terms;* that is, a quantity which differs less, than by any assignable quantity, from the value of the whole. This is also called the *limit* of the series. Thus the decimal 0.33333 &c. may come infinitely near to the vulgar fraction $\frac{1}{3}$, but can never exceed it, nor indeed exactly equal it. See arts. 453, 4. Therefore $\frac{1}{3}$ is the limit of 0.33333 &c. that is, of the series

$$\tfrac{3}{10}, \ \tfrac{3}{100}, \ \tfrac{3}{1000}, \ \tfrac{3}{10000}, \ \tfrac{3}{100000} \ \&c.$$

If the number of terms be supposed infinitely great, the

difference between their sum and $\frac{1}{3}$, will be infinitely small.

492. The sum of an infinite series whose terms decrease by a common divisor, may be found, by the rule for the sum of a series in geometrical progression. (Art. 442.) According to this, $S=\frac{rz-a}{r-1}$, that is, the sum of the series is found, by multiplying the greatest term into the ratio, subtracting the least term, and dividing by the ratio less 1. But, in an infinite series decreasing, the least term is infinitely small. It may be neglected, therefore, as of no comparative value. (Art. 456.) The formula will then become,

$$S=\frac{rz-0}{r-1} \quad \text{or} \quad S=\frac{rz}{r-1}.$$

Ex. 1. What is the sum of the infinite series

$$\tfrac{3}{10}, \tfrac{3}{100}, \tfrac{3}{1000}, \tfrac{3}{10000}, \tfrac{3}{100000}, \&c.$$

Here, the first term is $\frac{3}{10}$, and the ratio is 10.

Then $S=\dfrac{rz}{r-1}=\dfrac{10\times\frac{3}{10}}{10-1}=\dfrac{3}{9}=\dfrac{1}{3}$, the answer.

2. What is the sum of the infinite series

$$1+\tfrac{1}{2}+\tfrac{1}{4}+\tfrac{1}{8}+\tfrac{1}{16}+\tfrac{1}{32}+\tfrac{1}{64} \&c.$$

Ans. $S=\dfrac{rz}{r-1}=\dfrac{2\times 1}{2-1}=2.$

3. What is the sum of the infinite series

$$1+\tfrac{1}{3}+\tfrac{1}{9}+\tfrac{1}{27}+\tfrac{1}{81} \&c. \qquad \text{Ans. } \tfrac{3}{2}=1+\tfrac{1}{2}.$$

493. There are certain classes of infinite series, whose sums may be found by *subtraction*.

By the rules for the reduction and subtraction of fractions,

$$\frac{1}{2}-\frac{1}{3}=\frac{3-2}{2\times 3}=\frac{1}{2\times 3},$$

$$\frac{1}{3}-\frac{1}{4}=\frac{4-3}{3\times 4}=\frac{1}{3\times 4},$$

$$\frac{1}{4}-\frac{1}{5}=\frac{5-4}{4\times 5}=\frac{1}{4\times 5}, \&c.$$

If then the fractions on the right be formed into a series, they will be equal to the *difference* of two series formed from the fractions on the left. This difference is easily found: for if the first term be taken away from one of these two series, it will be equal to the other.

Suppose we have to find the sum of the infinite series

$$\frac{1}{2\cdot3}+\frac{1}{3\cdot4}+\frac{1}{4\cdot5}+\frac{1}{5\cdot6} \text{ \&c.}$$

From this, let another be derived, by removing the last factor from each of the denominators; and let the sum of the new series be represented by S,

That is, let $\quad S=\frac{1}{2}+\frac{1}{3}+\frac{1}{4}+\frac{1}{5}$ &c.

Then $\quad\quad S-\frac{1}{2}=\frac{1}{3}+\frac{1}{4}+\frac{1}{5}+\frac{1}{6}$ &c.

And by subtraction $\dfrac{1}{2}=\dfrac{1}{2\cdot3}+\dfrac{1}{3\cdot4}+\dfrac{1}{4\cdot5}+\dfrac{1}{5\cdot6}$ &c.

Here the new series is made one side of an equation, and directly under it, is written the same series, after the first term $\frac{1}{2}$ is taken away. If the upper one is equal to S, it is evident that the lower one must be equal to $S-\frac{1}{2}$. Then subtracting the terms of one equation from those of the other, (Ax. 2.) we have the sum of the proposed series equal to $\frac{1}{2}$. For $S-(S-\frac{1}{2})=S-S+\frac{1}{2}=\frac{1}{2}$.

2. What is the sum of the infinite series

$$\frac{1}{1\cdot3}+\frac{1}{2\cdot4}+\frac{1}{3\cdot5}+\frac{1}{4\cdot6}+\frac{1}{5\cdot7} \text{ \&c.}$$

Here a new series may be formed, as before, by omitting the last factor in each denominator.

Let $\quad\quad S=1+\frac{1}{2}+\frac{1}{3}+\frac{1}{4}+\frac{1}{5}$ &c.

Then $\quad S-\frac{3}{2}=\frac{1}{3}+\frac{1}{4}+\frac{1}{5}+\frac{1}{6}+\frac{1}{7}$ &c.

And by subtraction $\dfrac{3}{2}=\dfrac{2}{1\cdot3}+\dfrac{2}{2\cdot4}+\dfrac{2}{3\cdot5}+\dfrac{2}{4\cdot6}+\dfrac{2}{5\cdot7}$ &c.

Or $\quad\quad\quad \dfrac{3}{4}=\dfrac{1}{1\cdot3}+\dfrac{1}{2\cdot4}+\dfrac{1}{3\cdot5}+\dfrac{1}{4\cdot6}+\dfrac{1}{5\cdot7}$ &c.

In repeating the new series, in this case, it is necessary to omit the *two* first terms, which are $1 + \frac{1}{2} = \frac{3}{2}$.

3. What is the sum of the infinite series

$$\frac{1}{2 \cdot 4 \cdot 6} + \frac{1}{4 \cdot 6 \cdot 8} + \frac{1}{6 \cdot 8 \cdot 10} + \frac{1}{8 \cdot 10 \cdot 12} \text{ \&c.}$$

Here a new series may be formed, by omitting the last factor, and retaining the two first, in each denominator.

$$\text{Let } S = \frac{1}{2 \cdot 4} + \frac{1}{4 \cdot 6} + \frac{1}{6 \cdot 8} + \frac{1}{8 \cdot 10} \text{ \&c.}$$

$$\text{Then } S - \frac{1}{8} = \frac{1}{4 \cdot 6} + \frac{1}{6 \cdot 8} + \frac{1}{8 \cdot 10} + \frac{1}{10 \cdot 12} \text{ \&c.}$$

$$\text{And by subt. } \frac{1}{8} = \frac{4}{2 \cdot 4 \cdot 6} + \frac{4}{4 \cdot 6 \cdot 8} + \frac{4}{6 \cdot 8 \cdot 10} + \frac{4}{8 \cdot 10 \cdot 12} \text{ \&c.}$$

$$\text{Or } \frac{1}{32} = \frac{1}{2 \cdot 4 \cdot 6} + \frac{1}{4 \cdot 6 \cdot 8} + \frac{1}{6 \cdot 8 \cdot 10} + \frac{1}{8 \cdot 10 \cdot 12} \text{ \&c.}$$

4. What is the sum of the infinite series

$$\frac{1}{1 \cdot 2 \cdot 3} + \frac{1}{2 \cdot 3 \cdot 4} + \frac{1}{3 \cdot 4 \cdot 5} + \frac{1}{4 \cdot 5 \cdot 6} \text{ \&c.} \qquad \text{Ans. } \frac{1}{4} *$$

* See Note P.

COMPOSITION and RESOLUTION of the higher EQUATIONS.

Art. 494. **E**QUATIONS of any degree may be produced from *simple* equations, by multiplication. The manner in which they are compounded will be best understood, by taking them in that state in which they are all brought on one side by transposition. (Art. 178.) It will also be necessary to assign, to the same letter, different values, in the different simple equations.

Suppose, that in one equation, $x = 2$
And, that in another, $x = 3$
By transposition, $x - 2 = 0$
And $x - 3 = 0$

Mult. them together, $x^2 - 5x + 6 = 0$
Next, suppose $x - 4 = 0$

And multiplying, $x^3 - 9x^2 + 26x - 24 = 0$
Again, suppose $x - 5 = 0$

And mult. as before $x^4 - 14x^3 + 71x^2 - 154x + 120 = 0$ &c.

Collecting together the products, we have

$(x-2)(x-3)$ $= x^2 - 5x + 6 = 0$
$(x-2)(x-3)(x-4)$ $= x^3 - 9x^2 + 26x - 24 = 0$ [&c.
$(x-2)(x-3)(x-4)(x-5) = x^4 - 14x^3 + 71x^2 - 154x + 120 = 0$

That is,

The prod. of *two* simple equations, is a *quadratic* equation;
of *three* simple equations, is a *cubic* equation;
of *four* simple equations, is a *biquadratic*, or an equation of the fourth degree, &c. (Art. 300.)
Or, a cubic equation may be considered as the product

ef a quadratic and a simple equation: a biquadratic, as the product of two quadratic; or of a cubic and a simple equation, &c.

495. In each case, the *exponent* of the unknown quantity, in the first term, is equal to the degree of the equation; and, in the succeeding terms, it decreases regularly by 1, like the exponent of the leading quantity in the power of a binomial. (Art. 468.)

In a quadratic equation, the exponents are   2, 1.
In a cubic equation,   3, 2, 1.
In a biquadratic,   4, 3, 2, 1, &c.

496. The *number* of terms, is greater by 1, than the degree of the equation, or the number of simple equations from which it is produced. For, besides the terms which contain the different powers of the unknown quantity, there is one which consists of *known* quantities only. The equation is here supposed to be *complete*. But if there are, in the partial products, terms which balance each other, these may *disappear* in the result. (Art. 110.)

497. Each of the values of the unknown quantity is called a *root* of the equation.

Thus, in the example above,

The roots of the quadratic equation are   3, 2.
         of the cubic equation   4, 3, 2.
         of the biquadratic   5, 4, 3, 2.

The term *root* is not to be understood in the same sense here, as in the preceding sections. The root of an *equation* is not a quantity which multiplied into *itself* will produce the equation. It is one of the values of the unknown quantity; and when its sign is changed by transposition, it is a term in one of the binomial factors which enter into the composition of the equation of which it is a root.

The value of the unknown letter x, in the equation, is a quantity which may be substituted for x, without affecting the equality of the members. In the equations which we are now considering, each member is equal to 0; and the first is the product of several factors. This product will continue to be equal to 0, as long as any one of its factors is 0. (Art. 112.) If then in the equation

$$(x-2)\times(x-3)\times(x-4)\times(x-5)=0,$$

we substitute 2 for x, in the first factor, we have

$$0\times(x-3)\times(x-4)\times(x-5)=0.$$

So if we substitute 3 for x, in the second factor, or 4 in the third, or 5 in the fourth, the whole product will still be 0. This will also be the case, when the product is formed by an actual multiplication of the several factors into each other.

Thus, as $x^3 - 9x^2 + 26x - 24 = 0$; (Art. 494.)

So $2^3 - 9 \times 2^2 + 26 \times 2 - 24 = 0$,

And $3^3 - 9 \times 3^2 + 26 \times 3 - 24 = 0$, &c.

Either of these values of x, therefore, will satisfy the conditions of the equation.

498. The *number* of roots, then, which belong to any equation, is equal to the *degree* of the equation.

Thus, a quadratic equation has *two* roots;

a cubic equation, *three;*

a biquadratic, *four*, &c.

Some of these roots, however, may be *imaginary*. For an imaginary expression may be one of the factors from which the equation is derived.

499. The *resolution* of equations, which consists in finding their *roots*, cannot be well understood, without bringing into view a number of principles, derived from the manner in which the equations are compounded. The laws by which the *co-efficients* are governed, may be seen, from the following view of the multiplication of the factors

$$x-a,\ x-b,\ x-c,\ x-d,$$

each of which is supposed equal to 0.

The several co-efficients of the same power of x, are placed *under* each other.

Thus, $-ax - bx$ is written $\left.\begin{array}{c} -a \\ -b \end{array}\right\} x$; and the other co-efficients, in the same manner.

The product, then

Of $(x-a) = 0$.

Into $(x-b) = 0$

Is $x^2 \left.\begin{array}{c} -a \\ -b \end{array}\right\} x + ab = 0$, a quadratic equation.

This into $x - c = 0$

Is $x^3 \left.\begin{array}{c} -a \\ -b \\ -c \end{array}\right\} x^2 \left.\begin{array}{c} +ab \\ +ac \\ +bc \end{array}\right\} x - abc = 0$, a cubic equation.

This into $x - d = 0$

$$\text{Is } x^4 \begin{Bmatrix} -a \\ -b \\ -c \\ -d \end{Bmatrix} x^3 \begin{Bmatrix} +ab \\ +ac \\ +ad \\ +bc \\ +bd \\ +cd \end{Bmatrix} x^2 \begin{Bmatrix} -abc \\ -abd \\ -acd \\ -bcd \end{Bmatrix} x + abcd = 0, \text{ a biquadratic.}$$

&c.

500. By attending to these equations, it will be seen that,
In the *first* term of each, the co-efficient of x is 1:

In the *second* term, the co-efficient is the sum of all the roots of the equation, with contrary signs. Thus the roots of the quadratic equation are a and b, and the co-efficients, in the second term, are $-a$ and $-b$.

In the *third* term, the co-efficient of x is the sum of all the products which can be made, by multiplying together any *two* of the roots. Thus, in the cubic equation, as the roots are a, b, and c, the co-efficients, in the third term, are ab, ac, bc.

In the *fourth* term, the co-efficient of x is the sum of all the products which can be made, by multiplying together any *three* of the roots, after their signs are changed. Thus the roots of the biquadratic equation are a, b, c, and d, and the co-efficients in the fourth term, are $-abc$, $-abd$, $-acd$, $-bcd$.

The *last term* is the product formed from *all* the roots of the equation, after the signs are changed.

In the cubic equation, it is $-a \times -b \times -c = -abc$.
In the biquadratic, $-a \times -b \times -c \times -d = +abcd$, &c.

501. In the preceding examples, the roots are all *positive*. The signs are changed by transposition, and when the several factors are multiplied together, the terms in the product, as in the power of a residual quantity, (Art. 476.) are alternately positive and negative. But if the roots are all *negative*, they become positive by transposition, and *all* the terms in the product must be positive. Thus, if the several values of x are $-a$, $-b$, $-c$, $-d$, then

$$x+a=0, \ x+b=0, \ x+c=0, \ x+d=0;$$

and by multiplying these together, we shall obtain the same equations as before, except that the signs of all the terms will be positive. In other cases, some of the roots may be positive, and some of them negative.

Ii

502. As equations are raised, from a lower degree to a higher, by multiplication, so they may be *depressed*, from a higher degree to a lower, by *division*. The product of $(x-a)$ into $(x-b)$ is a quadratic equation; this into $(x-c)$ is a cubic equation; and this into $(x-d)$ is a biquadratic. (Art 494.) If we *reverse* this process, and divide the biquadratic by $(x-d)$, the quotient, it is evident, will be a cubic equation; and if we divide this by $(x-c)$, the quotient will be quadratic, &c. The divisor is one of the factors from which the equation is produced, that is, it is a binomial consisting of x and one of the roots with its sign changed. When, therefore, we have found either of the roots, we may divide by this, connected with the unknown quantity, which will reduce the equation to the next inferior degree.

503. Various methods have been devised for the *resolution* of the higher equations; but many of them are intricate and tedious, and others are applicable to particular cases only. The roots may be found, however, with sufficient exactness, by successive *approximations*. From the laws of the co-efficients, as stated in art. 500, a general estimate may be formed of the values of the roots. They must be such, that, when their signs are changed, their *product* shall be equal to the *last* term of the equation, and their *sum* equal to the coefficient of the *second* term. A trial may then be made, by substituting, in the place of the unknown letter, its supposed value. If this proves to be too small or too large, it may be increased or diminished, and the trials repeated, till one is found which will nearly satisfy the conditions of the equation. After we have discovered or assumed two approximate values, and calculated the errours which result from them, we may obtain a more exact correction of the root, by the following

Proportion.

As the difference of the errours, to the difference of the assumed numbers :

*So is the least errour, to the correction required in the corresponding assumed number.**

This is founded on the supposition, that the errours in the *results* are proportioned to the errours in the *assumed numbers.*

* See Hutton's Mathematics.

Let N and n be the assumed numbers;
    S and s,   the errours of these numbers;
    R and r,   the errours in the results.

Then by the supposition $R:r::S:s$
And subt. the consequents (Art.389.) $R-r:S-s::r:s$.

But the difference of the assumed numbers is the same, as the difference of their errours. If, for instance, the true number is 10, and the assumed numbers 12 and 15, the errours are 2 and 5; and the difference between 2 and 5 is the same, as between 12 and 15. Substituting, then, $N-n$ for $S-s$, we have $R-r:N-n::r:s$, which is the proportion stated above.

The term *difference* is to be understood here, as it is commonly used in algebra, to express the result of subtraction according to the general rule. (Art. 82.) In this sense, the difference of two numbers one of which is positive and the other negative, is the same, as their *sum* would be, if their signs were alike. (Art. 85.)

The supposition which is made the foundation of the rule for finding the true value of the root of an equation, is not strictly correct. The errours in the results are not *exactly* proportioned to the errours in the assumed numbers. But as a greater errour in the assumed number, will generally lead to a greater errour in the result, than a less one, the rule will answer the purpose of approximation. If the value which is first found, is not sufficiently correct, this may be taken as one of the numbers for a second trial; and the process may be repeated, till the errour is diminished as much as is required. There will generally be an advantage in assuming two numbers whose difference is 1, or .01, or .001, &c.

Ex. 1. Find the value of x, in the cubic equation.
$$x^3 - 8x^2 + 17x - 10 = 0.$$

Here, as the signs of the terms are alternately positive and negative, the roots must be all positive; (Art. 501.) their product must be 10, and their sum 8.

Let it be supposed that one of them is 5·1 or 5·2. Then, substituting these numbers for x, in the given equation, we have,

By the 1st suppos'n, $(5{\cdot}1)^3-8\times(5{\cdot}1)^2+17\times(5{\cdot}1)-10=1{\cdot}271,$
By the second, $\quad(5{\cdot}2)^3-8\times(5{\cdot}2)^2+17\times(5{\cdot}2)-10=2{\cdot}688,$

That is, By the first supposition. By the second supposition,

		By the first supp.	By the second supp.
The 1st term,	$x^3=$	$132{\cdot}651$	$140{\cdot}608$
The 2d,	$-8x^2=$	$-208{\cdot}08$	$-216{\cdot}32$
The 3d,	$17x=$	$86{\cdot}7$	$88{\cdot}4$
The 4th,	$-10=$	$-10{\cdot}$	$-10{\cdot}$

Sums or errours, $+1{\cdot}271$ $+2{\cdot}688$
Subtracting one from the other, $1{\cdot}271$

Their difference is $1{\cdot}417$

Then, stating the proportion,

$1{\cdot}4:0{\cdot}1::1{\cdot}27:0{\cdot}09,$ the correction to be sub-tracted from the first assumed number $5{\cdot}1$: The remainder is $5{\cdot}01$, which is a near value of x,

To correct this farther, assume $x=5{\cdot}01$, or $5{\cdot}02$

		By the first supposition.	By the second supposition.
The 1st term	$x^3=$	$125{\cdot}751$	$126{\cdot}506$
Tho 2d	$8x^2=$	$-200{\cdot}0$	$-201{\cdot}0$
The 3d	$17x=$	$85{\cdot}17$	$85{\cdot}34$
The 4th	$-10=$	$-10{\cdot}$	$-10{\cdot}$
Errours		$+0{\cdot}121$	$+0{\cdot}246$
			$0{\cdot}121$
Difference			$0{\cdot}125$

Then $0{\cdot}125:0{\cdot}01::0{\cdot}121:0{\cdot}01,$ the correction. This subtracted from $5{\cdot}01$, leaves 5 for the value of x; which will be found, on trial, to satisfy the conditions of the equation.

For $5^3-8\times5^2+17\times5-10=0.$

We have thus obtained one of the three roots. To find the other two, let the equation be divided by $x-5$, accor-ding to art. 462, and it will be depressed to the next inferi-our degree. (Art. 502.)

$x-5)x^3-8x^2+17x-10(x^2-3x+2=0,$

Here, the equation becomes quadratic.

By transposition, $\qquad x^2-3x=-2$
Completing the square; (Art.305.) $x^2-3x+\frac{9}{4}=\frac{9}{4}-2=\frac{1}{4}$
Extract. and transp. (Art. 303.) $x=\frac{3}{2}\pm\sqrt{\frac{1}{4}}=\frac{3}{2}\pm\frac{1}{2}.$

The first of these values of x is 2 and the other 1.

We have now found the three roots of the proposed equation. When their signs are changed, their sum is -8, the co-efficient of the second term, and their product -10 the last term.

2. What are the roots of the equation,

$\qquad x^3-8x^2+4x+48=0?$ $\qquad$ Ans. $-2,+4,+6.$

3. What are the roots of the equation

$\qquad x^3-16x^2+65x-50=0?$ $\qquad$ Ans. 1, 5, 10.*

$\qquad$ * See Note Q.

APPLICATION of ALGEBRA to GEOMETRY.

ART. 504. IT is often expedient to make use of the algebraic notation, for expressing the relations of geometrical quantities, and to throw the several steps in a demonstration into the form of equations. By this, the nature of the reasoning is not altered. It is only translated into a different *language*. *Signs* are substituted for *words*, but they are intended to convey the same meaning. A great part of the demonstrations in Euclid, really consist of a series of equations, though they may not be presented to us under the algebraic forms. Thus the proposition, that *the sum of the three angles of a triangle is equal to two right angles*, (Euc. 32. 1.) may be demonstrated, either in common language, or by means of the signs used in algebra.

Let the side *AB*, of the triangle *ABC*, (Fig. 1.) be continued to *D*; let the line *BE* be parallel to *AC*; and let *GHI* be a right angle.

The demonstration, in words, is as follows.

1. The angle *EBD* is *equal* to the angle *BAC*. (Euc. 29.1.)
2. The angle *CBE* is *equal* to the angle *ACB*.
3. Therefore, the angle *EBD added* to *CBE*, that is, the angle *CBD*, is *equal* to *BAC added* to *ACB*.
4. If to these equals, we add the angle *ABC*, the angle *CBD added* to *ABC*, is *equal* to *BAC added* to *ACB and ABC*.
5. But *CBD added* to *ABC*, is *equal* to twice *GHI*, that is, to two right angles. Euc. 13. 1.
6. Therefore, the angles *BAC, and ACB, and ABC*, are together equal to twice *GHI*, or two right angles.

* This and the following section are to be read *after* the Elements of Geometry.

Now, by substituting the sign +, for the word *added* or *and*, and the character =, for the word *equal*, we shall have the same demonstration, in the following form.

1. By Euclid 29.1. $EBD=BAC$
2. And $CBE=ACB$
3. Add. the two equa's, $EBD+CBE=BAC+ACB$
4. Ad.$ABC$to both sid's $CBD+ABC=BAC+ACB+ABC$
5. But, by Euc. 13.1, $CBD+ABC=2GHI$
6. Mak.the4th&5th equ.$BAC+ACB+ABC=2GHI.$

By comparing, one by one, the steps of these two demonstrations, it will be seen, that they are precisely the same, except that they are differently expressed. The algebraic mode has often the advantage, not only in being more *concise* than the other, but in exhibiting the *order* of the quantities more distinctly to the eye. Thus, in the fourth and fifth steps of the preceding example, as the parts to be compared are placed one under the other, it is seen, at once, what must be the new equation derived from these two. This regular arrangement is very important, when the demonstration of a theorem, or the resolution of a problem, is unusually complicated. In ordinary language, the numerous relations of the quantities require a series of explanations to make them understood; while, by the algebraic notation, the whole may be placed distinctly before us, at a single view. The disposition of the men on a chess-board, or the situation of the objects in a landscape, may be better comprehended, by a glance of the eye, than by the most laboured description in words.

505. It will be observed, that the notation in the example just given differs, in one respect, from that which is generally used in algebra. Each quantity is represented, not by a *single letter*, but by *several*. In common algebra, when one letter stands immediately before another, as *ab*, without any character between them, they are to be considered as *multiplied* together.

But, in geometry, AB is an expression for a *single line*, and not for the product of A into B. Multiplication is denoted, either by a point, or by the character $\times$. The product of AB into CD, is $AB \cdot CD$, or $AB \times CD$.

506. There is no impropriety, however, in representing a geometrical quantity by a single letter. We may make b stand for a line or an angle, as well as for a number.

If, in the example above, we put the angle

$$EBD=a, \qquad ACB=d, \qquad ABC=h,$$
$$BAC=b, \qquad CBD=g, \qquad GHI=l;$$
$$CBE=c,$$

the demonstration will stand thus,

1. By Euc. 29. 1. $a=b$
2. And $c=d$
3. Adding the two equations, $a+c=g=b+d$
4. Adding h to both sides, $g+h=b+d+h$
5. By Euc. 13. 1. $g+h=2l$
6. Making the 4th and 5th equal, $b+d+h=2l.$

This notation is, apparently, more simple than the other; but it deprives us of what is of great importance in geometrical demonstrations, a continual and easy reference to the figure. To distinguish the two methods, *capitals* are generally used, for that which is peculiar to geometry; and *small letters*, for that which is properly algebraic. The latter has the advantage, in long and complicated processes, but the other is often to be preferred, on account of the facility with which the figures are consulted.

507. If a line, whose length is measured from a given point or line, be considered *positive;* a line proceeding in the *opposite* direction is to be considered *negative.* If *AB*, (Fig.2.) reckoned from *DE* on the *right*, is positive; *AC* on the *left* is negative.

A line may be conceived to be produced by the *motion of a point.* Suppose a point to move in the direction of *AB*, and to describe a line varying in length with the distance of the point from *A.* While the point is moving towards *B*, its distance from *A* will *increase.* But if it move from *B* towards *C*, its distance from *A* will *diminish*, till it is reduced to nothing, and will then increase on the *opposite side.* As that which increases the distance on the right, diminishes it on the left, the one is considered positive, and the other negative. See arts. 59, 60.

Hence, if in the course of a calculation, the algebraic value of a line is found to be *negative;* it must be measured in a direction to opposite that which, in the same process, has been considered positive. (Art. 197.)

508. In algebraic calculations, there is frequent occasion for *multiplication, division,* involution, &c. But how, it may

be asked, can *geometrical* quantities be multiplied into each other. One of the factors, in multiplication, is always to be considered as a *number*. (Art. 91.) The operation consists in repeating the multiplicand, as many times as there are *units* in the multiplier. How then can a *line*, a *surface*, or a *solid*, become a multiplier?

To explain this, it will be necessary to observe, that whenever one geometrical quantity is multiplied into another, some *particular extent* is to be considered *the unit*. It is immaterial what this extent is, provided it remain the same, in different parts of the same calculation. It may be an inch, a foot, a rod, or a mile. If an *inch* is taken for the unit, each of the lines to be multiplied, is to be considered as made up of so many parts, as it contains inches. The multiplicand will then be repeated, as many times, as there are units in the multiplier. If, for instance, one of the lines be a foot long, and the other, half a foot; the factors will be, one 12 inches, and the other 6, and the product will be 72 inches. Though it would be absurd, to say that one line is to be repeated, *as often as another is long;* yet there is no impropriety in saying, that one is to be repeated as many times, as there are feet or rods in the other. This, the nature of a calculation often requires. ·

509. If the line which is to be the multiplier, is only a *part* of the length taken for the unit; the product is a like part of the multiplicand. (Art. 90.) Thus, if one of the factors is 6 inches, and the other half an inch, the product is 3 inches.

510. Instead of referring to the measures in common use, as inches, feet, &c. it is often convenient to fix upon one of the lines in a figure, as the unit with which to compare all the others. When there are a number of lines drawn within and about a *circle*, the *radius* is commonly taken for the unit. This is particularly the case in trigonometrical calculations.

511. The observations which have been made concerning lines, may be applied to *surfaces* and *solids*. There may be occasion to multiply the *area* of a figure, by the number of inches in some given line.

But here, another difficulty presents itself. The product of two lines is often spoken of, as being equal to a *surface;* and the product of a line and a surface, as equal to a *solid.* Thus the area of a parallelogram is said to be equal to the product of its base and height; and the solid contents of a

J j

cylinder, is said to be equal to the product of its length, into the area of one of its ends. But if a line has no *breadth*, how can the multiplication, that is, the *repetition*, of a line produce a surface? And if a surface has no *thickness*, how can a repetition of it produce a solid?

If a parallelogram, represented on a reduced scale by *ABCD*, (Fig. 3.) be five inches long, and three inches wide; the area or surface is said to be equal to the product of 5 into 3, that is, to the number of inches in *AB*, multiplied by the number in *BC*. But the inches in the lines *AB* and *BC* are *linear* inches, that is, inches in *length* only; while those which compose the surface *AC* are *superficial* or *square* inches, a different species of magnitude. How can one of these be converted into the other by multiplication, a process which consists in repeating quantities, without changing their nature?

512. In answering these inquiries, it must be admitted, that measures of length do not belong to the same class of magnitudes with superficial or solid measures; and that none of the steps of a calculation can, properly speaking, transform the one into the other. But, though a line can not become a surface or a solid, yet the several measuring units in common use are so adapted to each other, that squares, cubes, &c. are bounded by lines of the same name. Thus the side of a square inch, is a linear inch; that of a square rod, a linear rod, &c. The *length* of a linear inch is therefore, the same, as the length or breadth of a square inch.

If then, several square inches are placed together, as from *Q* to *R*, (Fig. 3.) the *number* of them in the parallelogram *OR* is the same, as the number of linear inches in the side *QR*: and, if we know the length of this, we have of course the area of the parallelogram, which is here supposed to be one inch wide.

But, if the breadth is *several* inches, the larger parallelogram contains as many smaller ones, each an inch wide, as there are inches in the whole breadth. Thus, if the parallelogram *AC* (Fig. 3.) is 5 inches long, and 3 inches broad, it may be divided into three such parallelograms as *OR*. To obtain then the number of squares in the large parallelogram, we have only to multiply the number of squares in one of the small parallelograms, into the number of such parallelograms contained in the whole figure. But the number of square inches in one of the small parallelograms, is equal to the number of linear inches in the *length AB*. And the number of small parallelograms, is equal to the number of linear inches in the *breadth BC*. It is therefore said concisely, that the *area of the parallelogram is equal to the length multiplied into the breadth.*

513. We hence obtain a convenient algebraic expression for the area of a right angled parallelogram. If two of the sides perpendicular to each other are *AB* and *BC*, the expression for the area is $AB \times BC$; that is, putting a for the area,

$$a = \dot{A}B \times BC.$$

It must be understood, however, that when *AB* stands for a *line*, it contains only *linear* measuring units; but when it enters into the expression for the *area*, it is supposed to contain *superficial* units of the same name. Yet as, in a given length, the *number* of one is equal to that of the other, they may be represented by the same letters, without leading to errour in calculation.

514. The expression for the area may be derived, by a method more simple, but less satisfactory perhaps to some, from the principles which have been stated concerning *variable quantities*, in the 13th section. Let a (Fig. 4.) represent a square inch, foot, rod, or other measuring unit; and let b and l be two of its sides. Also, let A be the area of any right angled parallelogram, B its breadth, and L its length. Then it is evident, that, if the breadth of each were the same, the areas would be as the lengths; and, if the length of each were the same, the areas would be as the breadths.

That is, $A : a :: L : l$, when the breadth is given,
And $A : a :: B : b$, when the length is given;
Therefore, (Art. 420.) $A : a :: B \times L : bl$, when both vary.

That is, the area is as the *product* of the *length* and *breadth*.

515. Hence, in quoting the Elements of Euclid, the term *product* is frequently substituted for *rectangle*. And whatever is there proved concerning the equality of certain rectangles, may be applied to the products of the lines which contain the rectangles.[*]

516. The area of an *oblique* parallelogram is also obtained, by multiplying the base into the perpendicular height. Thus the expression for the area of the parallelogram

[*] See Note R.

$ABNM$ (Fig. 5.) is $MN \times AD$, or $AB \times BC$. For, by art. 513, $AB \times BC$ is the area of the right angled parallelogram $ABCD$; and by Euclid 36.1, parallelograms upon equal bases, and between the same parallels, are equal; that is, $ABCD$ is equal to $ABNM$.

517. The area of a *square* is obtained, by multiplying one of the sides *into itself*. Thus the expression for the area of the square AC, (Fig. 6.) is $\overline{AB}^2$, that is,

$$a = \overline{AB}^2.$$

For the area is equal to $AB \times BC$. (Art. 513.)

But $AB = BC$, therefore, $AB \times BC = \overline{AB} \times AB = \overline{AB}^2$.

518. The area of a *triangle* is equal to *half* the product of the base and height. Thus the area of the triangle ABG, (Fig. 7.) is equal to half AB into GH or its equal BC, that is,

$$a = \tfrac{1}{2} AB \times BC.$$

For the area of the parallelogram $ABCD$ is $AB \times BC$. (Art. 513.) And, by Euc. 41. 1, if a parallelogram and a triangle are upon the same base, and between the same parallels, the triangle is half the parallelogram.

519. Hence, an algebraic expression may be obtained, for the area of any figure whatever which is bounded by right lines. For every such figure may be divided into triangles. Thus the right-lined figure

$ABCDE$ (Fig 8th.) is composed of the triangles ABC, ACE, and ECD.

The area of the triangle $ABC = \tfrac{1}{2} AC \times BL$,
That of the triangle $ACE = \tfrac{1}{2} AC \times EH$,
That of the triangle $ECD = \tfrac{1}{2} EC \times DG$.

The area of the whole figure is, therefore, equal to

$$(\tfrac{1}{2} AC \times BL) + (\tfrac{1}{2} AC \times EH) + (\tfrac{1}{2} EC \times DG).$$

The explanations, in the preceding articles, contain the first principles of the *mensuration of superficies*. The object of introducing the subject in this place, however, is not to make a practical application of it, at present; but merely to show the grounds of the method of representing geometrical quantities in algebraic language.

520. The expression for the superficies has here been de-

rived from that of a line or lines. It is frequently necessary to *reverse* this order; to find a side of a figure, from knowing its area.

If the number of square inches in the parallelogram *ABCD* (Fig. 3.) whose breadth *BC* is 3 inches, be divided by 3; the quotient will be a parallelogram *ABEF*, one inch wide, and of the same length with the larger one. But the length of the small parallelogram, is the length of its side *AB*. The number of *square* inches in one is the same, as the number of *linear* inches in the other. (Art. 512.) If therefore, the area of the large parallelogram be represented

by *a*, the side $AB = \dfrac{a}{BC}$, that is, *the length of a parallelogram is found, by dividing the area by the breadth.*

521. If *a* be put for the area of a *square* whose side is *AB*

Then by art. 517. $a = \overline{AB}^2$

And extracting both sides, $\sqrt{a} = AB$

That is, *the side of a square is found, by extracting the square root of the number of measuring units in its area.*

522. If *AB* be the base of a *triangle*, and *BC* its perpendicular height;

Then, by art. 518, $a = \frac{1}{2}BC \times AB$

And dividing by $\frac{1}{2}BC$, $\dfrac{a}{\frac{1}{2}BC} = AB,$

That is, *the base of a triangle is found, by dividing the area by half the height.*

523. As a *surface* is expressed, by the product of its length and breadth; the contents of a *solid* may be expressed, by the product of its length, breadth, and depth. It is necessary to bear in mind, that the measuring unit of solids is a *cube;* and that the side of a cubic inch, is a square inch, the side of a cubic foot, a square foot, &c.

Let *ABCD* (Fig. 3.) represent the base of a parallelopiped, 5 inches long, 3 inches broad, and *one* inch deep. It is evident there must be as many *cubic* inches in the solid, as there are *square* inches in its base. And, as the product of the lines *AB* and *BC* gives the area of this base, it gives, of course, the contents of the solid. But suppose that the depth of the parallelopiped, instead of being *one* inch, is

four inches. Its contents must be four times as great. If,
then, the length be *AB*, the breadth *BC*, and the depth *CO*,
the expression for the solid contents will be,

$$AB \times BC \times CO.$$

524. By means of the algebraic notation, a geometrical
demonstration may often be rendered much more simple and
concise, than in ordinary language. The proposition, (Euc.
4. 2.)| that when a straight line is divided into two parts, the
square of the whole line is equal to the squares of the two
parts, together with twice the product of the parts, is demonstrated, by involving a binomial.

Let the side of a square be represented by *s*;
And let it be divided into two parts, *a* and *b*.

By the supposition, $s = a + b$
And, squaring both sides, $s^2 = a^2 + 2ab + b^2.$
Or, changing the order of the terms, $s^2 = a^2 + b^2 + 2ab.$

That is, s^2 the square of the whole line, is equal to a^2, and
b^2, the squares of the two parts, together with $2ab$, twice the
product of the parts.

525. The algebraic notation may also be applied, with
great advantage, to the solution of geometrical *problems*. In
doing this, it will be necessary, in the first place, to raise an
algebraic equation, from the geometrical relations of the
quantities given and required; and then, by the usual reductions, to find the value of the unknown quantity in this equation. See art. 192.

Prob. 1. Given the *base*, and the *sum* of the hypothenuse and perpendicular, of the right angled triangle, *ABC*,
(Fig. 9.) to find the perpendicular.

Let the base $AB = b$ ⎫
The perpendicular $BC = x$ ⎬
The sum of hyp. and perp. $x + AC = a$ ⎪
Then transposing *x*, $AC = a - x$ ⎭

1. By Euclid 47. 1, $\overline{BC}^2 + \overline{AB}^2 = \overline{AC}^2$
2. That is, by the notation, $x^2 + b^2 = (a-x)^2 = a^2 - 2ax + x^2.$

Here we have a common algebraic equation, containing
only one unknown quantity. The reduction of this equa-

tion, in the usual manner, will give the value of x, the side required.

3. Transp. and uniting terms, $\qquad 2ax = a^2 - b^2$

4. Dividing by $2a$, $\qquad\qquad x = \dfrac{a^2 - b^2}{2a} = BC.$

The solution, in letters, will be the same, for any right angled triangle whatever, and may be expressed in a general theorem, thus; 'In a right angled triangle, the perpendicular is equal to the square of the sum of the hypothenuse and perpendicular, diminished by the square of the base, and divided by twice the sum of the hypothenuse and perpendicular.'

It is applied to particular cases, by substituting *numbers*, for the letters a and b. Thus, if the base is 8 feet, and the sum of the hypothenuse and perpendicular 16, the expression $\dfrac{a^2 - b^2}{2a}$ becomes $\dfrac{16^2 - 8^2}{2 \times 16} = 6$, the perpendicular; and this subtracted from 16, the sum of the hypothenuse and perpendicular, leaves 10, the length of the hypothenuse.

To prove that the answer is correct, we have only to observe, that in conformity with Euclid 47. 1, $8^2 + 6^2 = 10^2$.

Prob. 2. Given the *base*, and the *difference* of the hypothenuse and perpendicular, of a right angled triangle, to find the perpendicular.

Let the base $\qquad AB$(Fig. 10.)$= b = 20$
The perpendicular, $\qquad BC = x$
The given difference, $\qquad = d = 10$
Then will the hypothenuse $AC = x + d.$ For the greater of two quantities, is equal to the less added to their difference. Then

1. By Euclid 47. 1, $\qquad\qquad \overline{AC}^2 = \overline{AB}^2 + \overline{BC}^2$
2. That is, by the notation, $\qquad (x+d)^2 = b^2 + x^2$
3. Expanding $(x+d)^2$, (Art.217.) $x^2 + 2dx + d^2 = b^2 + x^2$
4. Transp. and uniting terms, $\qquad 2dx = b^2 - d^2$
5. Dividing by $2d$, $\qquad\qquad x = \dfrac{b^2 - d^2}{2d} = 15.$

Prob. 3. If the hypothenuse of a right-angled triangle is 30 feet, and the difference of the other two sides 6 feet, what is the length of the base?   Ans. 24 feet.

Prob. 4. If the hypothenuse of a right angled triangle is 50 rods, and the base is to the perpendicular as 4 to 3, what is the length of the perpendicular?   Ans. 30

Prob. 5. Having the perimeter and the diagonal of a parallelogram $ABCD$, (Fig. 11.) to find the sides.

Let the diagonal $\qquad AC = h = 10$
The side $\qquad AB = x$
Half the perimeter $BC + AB = BC + x = b = 14$
Then, by transposing x, $\qquad BC = b - x$

1. By Euclid 47. 1, $\qquad \overline{AB}^2 + \overline{BC}^2 = \overline{AC}^2$
2. That is, $\qquad x^2 + (b-x)^2 = h^2$
3. Expanding $(b-x)^2$, $\qquad x^2 + b^2 - 2bx + x^2 = h^2$
4. Transp. unit. and divid. by 2, $x^2 - bx = \frac{1}{2}h^2 - \frac{1}{2}b^2$
5. Completing the square, $x^2 - bx + \frac{1}{4}b^2 = \frac{1}{4}b^2 + \frac{1}{2}h^2 - \frac{1}{2}b^2$
6. Extract. and transp. $\qquad x = \frac{1}{2}b \pm \sqrt{\frac{1}{4}b^2 + \frac{1}{2}h^2 - \frac{1}{2}b^2} = 8.$

Here the side AB is found; and the side BC is equal to $b - x = 14 - 8 = 6.$

Prob. 6. The area of a right angled triangle ABC (Fig. 12.) being given, and the sides of a parallelogram inscribed in it, to find the side BC.

Let the given area $= a$, $\qquad DE = BF = b$,
$EB = DF = d$, $\qquad BC = x.$
Then by the figure, $\qquad CF = BC - BF = x - b.$
1. By similar triangles, $CF : DF :: BC : AB$
2. That is, $\qquad x - b : d :: x : AB$
3. Mult. ext. and means, $dx = (x-b) \times AB$
4. By art. 518, $\qquad a = AB \times \frac{1}{2}BC = AB \times \frac{1}{2}x$
5. Dividing by $\frac{1}{2}x$, $\qquad \dfrac{2a}{x} = AB$
6. Subs. in the 3d, $\dfrac{2a}{x}$ for AB, $dx = (x-b) \times \dfrac{2a}{x} = 2a - \dfrac{2ab}{x}$
7. Mult. by x, and trans. $dx^2 - 2ax = -2ab$
8. Dividing by d, $\quad x^2 - \dfrac{2ax}{d} = -\dfrac{2ab}{d}$
9. Complet. the square $x^2 - \dfrac{2ax}{d} + \dfrac{a^2}{d^2} = \dfrac{a^2}{d^2} - \dfrac{2ab}{d}$
10. Extract. and transp. $x = \dfrac{a}{d} + \sqrt{\dfrac{a^2}{d^2} - \dfrac{2ab}{d}} = BC.$

Prob. 7. The three sides of a right angled triangle ABC, (Fig. 13.) being given, to find the segments made by a perpendicular, drawn from the right angle to the hypothenuse.

The perpendicular will divide the original triangle, into two right angled triangles, BCD and ABD. (Euc. 8. 6.)

1. By Euc. 47. 1; $$\overline{BD}^2 + \overline{CD}^2 = \overline{BC}^2$$

2. By the figure, $$CD = AC - AD$$

3. Squar. both sides, $$\overline{CD}^2 = (AC - AD)^2$$

4. Therefore, $$\overline{BD}^2 + (AC - AD)^2 = \overline{BC}^2$$

5. Expanding, $$\overline{BD}^2 + \overline{AC}^2 - 2AC.AD + \overline{AD}^2 = \overline{BC}^2$$

6. Transposing, $$\overline{BD}^2 = \overline{BC}^2 - \overline{AC}^2 + 2AC.AD - \overline{AD}^2$$

7. By Euc. 47. 1, $$\overline{BD}^2 = \overline{AB}^2 - \overline{AD}^2$$

8. Mak. 6th and 7th eq. $$\overline{BC}^2 - \overline{AC}^2 + 2AC.AD = \overline{AB}^2$$

9. Transposing, $$2AC.AD = \overline{AB}^2 + \overline{AC}^2 - \overline{BC}^2$$

10. Dividing by $2AC$, $$AD = \frac{\overline{AB}^2 + \overline{AC}^2 - \overline{BC}^2}{2AC}$$

The *unknown* lines, to distinguish them from those which are known, are here expressed by Roman letters.

Prob. 8. Having the area of a parallelogram $DEFG$ (Fig. 14.) inscribed in a given triangle, ABC, to find the sides of the parallelogram.

Draw CI perpendicular to AB. By supposition, DG is parallel to AB. Therefore,

The triangle CHG, is similar to CIB ⎫
And CDG,   to CAB ⎬

K k

Let $CI=d$, $DG=x$,
$AB=b$, The given area $=a$.

1. By sim. trian.	$CB:CG::AB:DG$
2. And	$CB:CG::CI:CH$
3. By eq. ratios, (Art. 384.)	$AB:DG::CI:CH$
4. Mult. ext. and means,	$\dfrac{DG \times CI}{AB}=CH$
5. By the figure,	$CI-CH=IH=DE$
6. Substitut. for CH,	$CI-\dfrac{DG \times CI}{AB}=DE$
7. That is,	$d-\dfrac{dx}{b}=DE$
8. By art. 513,	$a=DG \times DE=x \times \left(d-\dfrac{dx}{b}\right)$
9. That is,	$a=dx-\dfrac{dx^2}{b}$
10. Transp. and mult. by b,	$dx^2-bdx=-ab$
11. Dividing by d,	$x^2-bx=-\dfrac{ab}{d}$
12. Completing the square,	$x^2-bx+\dfrac{b^2}{4}=\dfrac{b^2}{4}-\dfrac{ab}{d}$
13. Extract. and transp.	$x=\dfrac{b}{2}-\sqrt{\dfrac{b^2}{4}-\dfrac{ab}{d}}=DG.$

The side DE is found, by dividing the area by DG.

Prob. 9. Through a given point, in a given circle, so to draw a right line, that its parts, between the point and the periphery, shall have a given difference.

In the circle $AQBR$, (Fig. 15.) let P be a given point, in the diameter AB.

Let $AP=a$, $PR=x$,
$BP=b$, The given difference $=d$,
Then will $PQ=x+d$.

1. By Euc. 35. 3 $PR \times PQ = AP \times BP$
2. That is, $x \times (x+d) = a \times b$
3. Or, $x^2 + dx = ab$
4. Completing the square, $x^2 + dx + \frac{1}{4}d^2 = \frac{1}{4}d^2 + ab$
5. Extract. and transp. $x = -\frac{1}{2}d \pm \sqrt{\frac{1}{4}d^2 + ab} = PR.$

With a little practice, the learner may very much abridge these solutions, and others of a similar nature, by reducing several steps to one.*

* See Note S.

EQUATIONS of CURVES.

ART. 526. IN the preceding section, algebra has been applied to geometrical figures, bounded by *right lines*. Its aid is required also, in investigating the nature and relations of *curves*. The advances which, in modern times, have been made in this department of geometry, are, in a great measure, owing to the method of expressing the distinguishing properties of the different kinds of lines, in *the form of equations*. To understand the principles on which inquiries of this sort are conducted, it is necessary to become familiar with the plan of notation which has been generally agreed upon.

527. *The positions of the several points in a curve drawn on a plane, are determined, by taking the distance of each from two right lines perpendicular to each other.*

Let the lines *AF* and *AG* (Fig. 16.) be perpendicular to each other. Also, let the lines *DB*, *D'B'*, *D"B"* be perpendicular to *AF*; and the lines *CD*, *C'D'*, *C"D"*, perpendicular to *AG*. Then the position of the point *D* is known, by the length of the lines *BD* and *CD*. In the same manner, the point *D'* is known, by the lines *B'D'* and *C'D'*; and the point *D"*, by the lines *B"D"* and *C"D"*. The two lines which are thus drawn, from any point in the curve, are, together, called the *co-ordinates* belonging to that point.

But, as there is frequent occasion to speak of each of the lines separately, one of them for distinction sake, is called an *ordinate*, and the other, an *abscissa*. Thus *BD* is the ordinate of the point *D*, and *CD*, or its equal *AB*, the abscissa of the same point. It is, generally, most convenient to take the abscissas on the line *AF*, as *AB* is equal to *CD*, *AB'* to *C'D'*, and *AB"* to *C"D"*. Euc. 33. 1. The lines *AF* and *AG*, to which the co-ordinates are drawn, are called the *axes* of the co-ordinates.

528. If co-ordinates could be drawn to *every* point in a curve, and, if the relations of the several abscissas to their

corresponding ordinates could be expressed by an equation; the position of each point, and consequently, the nature of the curve would be determined. Many important properties of the figure might also be discovered, merely by throwing the equation into different forms, by transposing, dividing, involving, &c. But the number of points in a line is unlimited. It is impossible, therefore, actually to draw co-ordinates to every one of them. Still there is a way in which an equation may be obtained, that shall be applicable to all the parts of a curve. This is effected, by making the equation depend on some property, which is *common to every pair of co-ordinates.* In explaining this, it will be proper to begin with a *straight line,* instead of a curve.

Let *AH* (Fig. 17.) be a line from which co-ordinates are drawn, on the axes *AF* and *AG* perpendicular to each other. And let the angle *FAH* be such, that the abscissa *CD* or *AB* shall be equal to *twice* the ordinate *BD*.

The triangles *ABD, AB'D', AB"D",* &c. are all similar. (Euc. 29. 1.) Therefore,

$$AB : BD :: AB' : B'D' :: AB'' : B''D'',$$

And if $AB = 2BD$, then $AB' = 2B'D'$, and $AB'' = 2B''D''$, &c.

That is, each abscissa is equal to twice the corresponding ordinate. But, instead of a separate equation for each pair of co-ordinates, one will be sufficient for the whole. Let x represent any one of the abscissas, and y, the ordinate belonging to the same point. Then,

$$x = 2y, \text{ or } y = \tfrac{1}{2}x,$$

This is a *general equation,* expressing the ratio of the co-ordinates of the line *AH* to each other. It differs from a common equation, in this, that x and y, have no determinate magnitude. The only condition which limits them is, that they shall be the abscissa and ordinate of the *same point.*

If $x = AB$, then $y = BD$
If $x = AB'$, $y = B'D'$
If $x = AB''$, $y = B''D''$, &c.

From this it is evident, that, if one of the co-ordinates be taken of any particular length, the other will be given by the equation. If, for instance, the abscissa x be two inches long, the ordinate y, which is half x, must be one inch.

If $x=8$, then $y=4$, If $x=30$, then $y=15$,
If $x=10$, $y=5$, If $x=100$, $y=50$, &c.

On the other hand, if $y=2$, then $x=4$, &c.

529. If the angle HAF be of any different magnitude, as in Fig. 18, the general equation will be the same, except the co-efficient of x. Let the ratio of y to x be expressed by a, that is, let $y:x::a:1$. Then by converting this into an equation, we have

$$ax=y.$$

The co-efficient a will be a whole number or a fraction, according as y is greater or less than x.

530. To apply these explanations to curves, let it be required to find a general equation of the common *parabola*. (Fig. 19.) It is the distinguishing property of this figure, as will be shown under Conic Sections, that the abscissas are proportioned to the *squares* of their ordinates. Let the ratio of the square of any one ordinate to its abscissa, be expressed by a. As the ratio is the same, between the square of any other ordinate of the parabola and its abscissa, we have universally $y^2:x::a:1$; and by converting this into an equation,

$$ax=y^2.$$

This is called the *equation of the curve*. The important advantages gained by this general expression, are owing to this, that the equation is equally applicable to *every point* of the curve. Any value whatever may be assigned to the abscissa x, provided the ordinate y is considered as belonging to the same point. But, while x and y vary together, the quantity a is supposed to remain constant.

By the equation of the parabola $ax=y^2$, and, extracting the root of both sides, (Art. 297.)

$$y=\sqrt{ax}. \text{ If } a=2, \text{ then } y=\sqrt{2x}. \text{ And}$$

If $x=4.5=AB$ (Fig.19.)then $y=\sqrt{2\times 4.5}=\sqrt{9}=3=BD$
If $x=8.\ =AB'$ $y=\sqrt{2\times 8}\ =\sqrt{16}=4=B'D'$
If $x=12.5=AB''$ $y=\sqrt{2\times 12.5}=\sqrt{25}=5=B''D''$
If $x=18.\ =AB'''$ $y=\sqrt{2\times 18}=\sqrt{36}=6=B'''D'''$.

531. When ordinates are drawn on *both sides* of the axis

tó which they are applied; those on one side will be *positive,* while those on the other side will be *negative.* Thus, in Fig. 19, if the ordinates on the *upper side* of *AF* be considered positive, those on the *under side* will be negative. (Art. 507.) The abscissas also are either positive or negative, according as they are on one side or the other of the point from which they are measured. Thus, in Fig. 20, if the abscissas on the right, *AB*, *AB'*, &c. be considered positive, those on the left, *AC*, *AC''*, &c. will be negative. And, in the solution of a problem, if an abscissa or an ordinate is found to be negative, it must be set off, on the side of the axis opposite to that on which the values are positive.

532. In the preceding instances, the straight line or curve to which the ordinates and abscissas are applied, crosses the axis, in the point where it is intersected by the other axis. Thus the curve (Fig. 19.) and the straight line *E'D'* (Fig. 20.) cross the axis *AF*, in the point *A*, where it is cut by the axis *AG*. But this is not always the case. The abscissas on the axis *QF* (Fig. 21.) may be reckoned from the line *GN*.

Let *x* represent any one of the abscissas, *MB*, *MB'*, &c. and *y* the corresponding ordinate.

$$\text{Let } z = AB, \qquad b = MA,$$
And *a* = the ratio of *BD* to *AB*, as before.

Then $az = y$, (Art. 529.) that is, $\qquad z = \dfrac{y}{a}$

But, by the figure, $AB = MB - MA$, i. e. $\qquad z = x - b$

Making the two equations equal; $\qquad x - b = \dfrac{y}{a}$

And transposing $-b$ $\qquad x = \dfrac{y}{a} + b.$

533. In investigating the properties of curves, it is important to be able to distinguish readily, the cases in which the abscissas or ordinates are *positive,* from those in which they are *negative;* and to determine, under what circumstances, either of the co-ordinates vanishes. *An abscissa vanishes at the point where the curve meets the axis from which the abscissas are measured.* And an ordinate vanishes, at the point where the curve meets the axis from which the ordinates are measured.

Thus, in Fig. 19, the ordinates are measured from the line

AF. The length of each ordinate is the *distance* of a particular point in the curve from the line. As the curve approaches the axis, the ordinate diminishes, till it becomes nothing, at the point of intersection. For, here, there is *no* distance between the curve and the axis.

The *abscissas* are measured from the line *AG*. These must diminish also, as the curve approaches this line, and become nothing at *A*.

534. From this it is evident, that, when the two axes meet the curve at the *same point*, the two co-ordinates *vanish together*. In Fig. 19, the two axes meet the curve at *A*, the one cutting, and the other touching it. But, in Fig. 21, the axis *MF* crosses the line *ND* at *A*; while *GN* crosses it at *N*. The ordinate, being the distance from *MF*, vanishes at *A*, where this distance is nothing. But the abscissa, being the distance from *GN*, vanishes at *N* or *M*.

535. An abscissa or an ordinate changes from positive to negative, by passing through the point where it is equal to 0. Thus the ordinate *y* (Fig. 20.) diminishes, as it approaches the point *A*; here it is nothing, and, on the other side of *A*, it becomes negative, because it is below the axis *CF*. (Art. 507.) In the same manner, the *abscissa*, on the right of *AG*, diminishes, as it approaches this line, becomes 0 at *A*, and then negative, on the left.

In this case, the two co-ordinates change from positive to negative, at the same point. But, in Fig. 21, the ordinates change from positive to negative at *A*; while the abscissas continue positive to *GN*, being still on the right of that line. On the right from *A*, the co-ordinates are both positive: between *A* and the line *GN*, the abscissas are positive, and the ordinates negative: and, on the left of *GN*, both are negative.

536. The most important applications of the principles stated in this section, will come under consideration, in succeeding branches of the mathematics, particularly in Fluxions. A few examples will be here given, to illustrate the observations which have now been made,

Prob. 1. To find the equation of the *circle*.

In the circle *FGM*, (Fig. 22.) let the two diameters *GN* and *FM* be perpendicular to each other. From any point in the curve, draw the ordinate *DB* perpendicular to *AF*; and *AB* will be the corresponding abscissa.

Let the radius $AD=r$, $\quad AB=x$, $\quad BD=y$.

Then, by Euc. 47. i, $\qquad \overline{BD}^2 = \overline{AD}^2 - \overline{AB}^2$

That is, $\qquad y^2 = r^2 - x^2$

And by evolution, $\qquad y = \pm \sqrt{r^2 - x^2}$

In the same manner, $\qquad x = \pm \sqrt{r^2 - y^2}$

That is, the abscissa is equal to the square root of the difference between the square of the radius and the square of the ordinate.

If the radius of the circle be taken for a *unit*, (Art. 510.) its square will also be 1, and the two last equations will become

$$y = \pm \sqrt{1-x^2}, \quad \text{and} \quad x = \pm \sqrt{1-y^2}$$

These equations will be the same, in whatever part of the arc GDF the point D is taken. For the co-ordinates will be the legs of a right angled triangle, the hypothenuse of which will be equal to AD, because it is the radius of the circle.

537. To understand the application to the other quarters of the circle, it must be observed, that, in each of the equations, the root is *ambiguous*. The values of y and of x may be either positive or negative. This results from the nature of a quadratic equation. (Art. 297.) It corresponds also with the situation of the different parts of the circle, with respect to the two diameters FM and GN. In the first quarter GF, the co-ordinates are supposed to be both positive. In the second, GM, the ordinates are still positive, but the abscissas become negative. (Art. 531.) In the third, MN, both are negative: and in the fourth, NF, the ordinates are negative, but the abscissas positive. That is,

In the quadrant $\begin{cases} FG, \; x \text{ is } +, \text{ and } y+, \\ GM, \; x \quad -, \qquad y+, \\ MN, \; x \quad -, \qquad y-, \\ NF, \; x \quad +, \qquad y-. \end{cases}$

538. In geometry, lines are supposed to be produced by

L l

the *motion of a point*. If the point moves uniformly in one direction, it produces a *straight* line. If it continually *varies* its direction, it produces a *curve*. The particular nature of the curve depends on certain conditions by which the motion is regulated. If, for instance, one point moves in such a manner, as to keep constantly at the same distance from another point which is fixed, the figure described is a *circle*, of which the fixed point is the centre. It is evident from the preceding problem, that the *equation* of this curve depends on the manner of description. For it is derived from the property, that different parts of the periphery are equally distant from the centre. In a similar manner, the equations of other curves may be derived from the law by which they are described; as will be seen in the following examples.

. Prob. 2. To find the equation of the curve called the *Cissoid* of Diocles. (Fig. 23.)

The description, which may be considered as the *definition* of the figure, is as follows.

In the diameter *AB*, of the semi-circle *ANB*, let the point *R* be at the same distance from *B*, as *P* is from *A*. Draw *RN* perpendicular to *AB*, to cut the circle in *N*. From *A*, through *N*, draw a straight line, extending if necessary beyond the circle. And from *P*, raise a perpendicular, to cut this line in *M*. The curve passes through the point *M*.

By taking *P* at different distances from *A*, as in Fig. 24, any number of points in the curve may be determined. As the line *PM* moves towards *B*, it becomes longer and longer; so as to extend the Cissoid beyond the semicircle.

To find the *equation* of the curve, let *AH* and *AB* be the axes of the co-ordinates.

Also, let each of the abscissas *AP*, *AP'*, *AP''*, &c. $= x$,
each of the ordinates *PM*, *P'M'*, *P''M''*, &c. $= y$,
and the diameter   *AB* $= b$,
Then, by the construction, $PB = AB - AP = b - x.$

As *PM* and *RN* are each perpendicular to *AB*, the triangles *APM* and *ARN* are similar. (Euc. 27 and 29. 1.) Therefore,

1. By sim. triangles, $\quad AP : PM :: AR : RN$

2. Or, by putting PB for its equal AR,
$$AP : PM :: PB : RN$$

3. Mult. ext. and means, $\quad \dfrac{PM \times PB}{AP} = RN$

4. Squaring both sides, $\quad \dfrac{\overline{PM}^2 \times \overline{PB}^2}{\overline{AP}^2} = \overline{RN}^2$

5. By Euc. 35. 3, and 3. 3, $\quad AR \times RB = \overline{RN}^2$

6. Or, putting PB for its equal AR, and AP for its equal RB,
$$PB \times AP = \overline{RN}^2$$

7. Mak. 4th and 6th equal, $\quad PB \times AP = \dfrac{\overline{PM}^2 \times \overline{PB}^2}{\overline{AP}^2}$

8. Mult. by $\overline{AP}^2$, and divid. by PB, $\quad \overline{AP}^3 = \overline{PM}^2 \times PB$

9. Or, $\qquad\qquad\qquad\qquad x^3 = y^2 \times (b - x)$

That is, the cube of the abscissa is equal to the square of the ordinate, multiplied by the difference between the diameter of the circle, and the abscissa. The equation is the same for every pair of co-ordinates.

Prob. 3. To find the equation of the *Conchoid* of Nicomedes.

To describe the curve, let AB Fig. 25. be a line given in position, and C a point without the line. About this point, let the line Ch revolve. From its intersections with AB, make the distances EM, $E'M'$, $E''M''$, &c. each equal to AD. The curve will pass through the points D, M, M', M'', &c.

To find its *equation*, let CD and AB be the axes of the co-ordinates. Draw FM parallel to AP, and PM parallel to CF. From the construction, AD is equal to EM.

Let the abscissa $\qquad AP = FM = x$,
the ordinate $\qquad PM = AF = y$,
the given line $\qquad CA = a$,
and $\qquad\qquad AD = EM = b$,
Then will $\qquad CF = CA + AF = a + y$.

As CM cuts the parallels CD and PM, and also the paral-lels AP and FM, the triangles CFM and MPE are similar. Then

1. By sim. triangles, $\qquad CF:FM::PM:PE$

2. Mult. ext. and means, $\qquad PE=\dfrac{FM\times PM}{CF}$

3. Squaring both sides, $\qquad \overline{PE}^2=\dfrac{\overline{FM}^2\times\overline{PM}^2}{\overline{CF}^2}$

4. By Euc. 47. 1, $\qquad \overline{PE}^2=\overline{EM}^2-PM^2$

5. Mak. 3d and 4th equal, $\quad \overline{EM}^2-\overline{PM}^2=\dfrac{\overline{FM}^2\times\overline{PM}^2}{\overline{CF}^2}$

6. That is, $\qquad\qquad\qquad b^2-y^2=\dfrac{x^2y^2}{(a+y)^2}$

7. Mult. by $(a+y)^2$ $\qquad (a+y)^2\times(b^2-y^2)=x^2y^2.$

539. In these examples, the equation is derived from the description of the curve. But this order may be reversed. If the equation is given, the curve may be described. For the equation expresses the relation of every abscissa to the corresponding ordinate. The curve is described, therefore, *by taking abscissas of different lengths, and applying ordinates to each.* The line required, will pass through the extremities of these ordinates.

Prob. 4. To describe the curve whose equation is

$$2x=y^2, \qquad\qquad \text{or} \quad y=\sqrt{2x}.$$

On the line AF, (Fig. 19.) take abscissas of different lengths :

For instance, $AB = 4\cdot5$, then the ordinate $BD=3$, (Art.530.)
$\qquad\qquad AB' = 8\cdot \qquad\qquad\qquad B'D' =4,$
$\qquad\qquad AB'' =12\cdot5 \qquad\qquad\qquad B''D'' =5,$
$\qquad\qquad AB''' =18\cdot \qquad\qquad\qquad B'''D''' =6,$
$\qquad\qquad\qquad\qquad \&c.$

Apply these several ordinates to their abscissas, and connect the extremities by the line $ADD'D''$, &c. which will be the curve required. The description will be more or less

accurate, according to the number of points for which ordinates are found.

540. If a point is conceived to move in such a manner, as
to pass through the extremities of all the ordinates assigned
by an equation; the line which it describes is called the *locus*
of the point, that is, the path in which it moves, and in which
it may always be found. The line is also called the *locus of*
the equation by which the successive positions of the point
are determined. Thus the common parabola (Fig. 19.) is
called the *locus* of the points D, D', D'', &c. or of the equation $ax=y^2$. (Art. 530.)   The arc of a circle is the *locus*
of the equation $x=\pm \sqrt{r^2 -x^2}$. (Art. 536.)   To find the
locus of an equation, therefore, is the same thing, as to find
the straight line or curve to which the equation belongs.

Prob. 5. To find the *locus* of the equation

$$x=\frac{y}{a}, \qquad\qquad \text{or} \quad ax=y,$$

in which, x and y are variable co-ordinates, while a is a determinate quantity.

If the abscissa x be taken of different lengths, the ordinate y must vary in such a manner as to preserve $ax=y$; or,
converting the equation into a proportion, $y:x::a:1$. Therefore, as a is a determinate quantity, the ratio of x to y will
be invariable; that is, any one abscissa will be to its ordinate,
as any other abscissa to its ordinate. Let two of the abscissas be AB and AB', (Fig. 17.) and their ordinates, BD and
$B'D'$; then,

$$AB:BD::AB':B'D'.$$

The line ADD' is, therefore, a *straight* line : (Euc. 32.6.)
and this is the *locus* of the equation.

If the proposed equation is $x=\frac{y}{a}+b$, the additional term
b, makes no difference in the nature of the *locus*. For the
only effect of b, is to lengthen the abscissas, so that they
must not be measured from A, but from some other point, as
M, (Fig. 21.) The ratio of AB, AB', &c. to BD', $B'D'$,&c.
still remains the same. See art. 532.   The *locus* of the
equation is, therefore, a straight line.

541. From this it will be easy to prove, that the *locus* of
every equation in which the co-ordinates x and y are in sepa-

rate terms, and do not rise above the *first power* is a straight line. For every such equation may be brought to the form $x = \frac{y}{a} \pm b$. All the terms may be reduced to three, one containing x, another y, and a third, the aggregate of the constant quantities which are not co-efficients of x and y; as will be seen, in the following problem.

Prob. 6. To find the *locus* of the equation

$$cx - d + hx - y + m = n.$$

By transposition, $\qquad cx + hx = y + n - m + d$

Dividing by $c + h$ $\qquad x = \frac{y}{c+h} + \frac{n-m+d}{c+h}$

Here, the constant quantities, in each term, may be represented by a single letter.(Art.321.) If, then, we make $c + h = a$, and $\frac{n-m+d}{c+h} = b$; the equation will become $x = \frac{y}{a} + b$, whose locus, by the last article, is a straight line.

542. But if the ordinates are as the squares, cubes, or higher powers of the abscissas, the *locus* of the equation, instead of being a straight line, is a curve. For the ordinates applied to a straight line, have the same ratio to each other which their abscissas have. But quantities have not the same ratio to each other, which their squares, cubes, or higher powers have. (Art. 354.) Thus, if $x^2 = y$, the ordinates will increase more rapidly than the abscissas. If the abscissas be taken, 1, 2, 3, 4, &c. the ordinates will be equal to their squares, 1, 4, 9, 16, &c.

543. As an unlimited variety of equations may be produced, by different combinations and powers of the co-ordinates, and as each of these has its appropriate *locus*; it is evident that the forms of curves must be innumerable. They may, however, be reduced to *classes*. The modern mode of classing them, is from the degree of their equations. *The different orders of lines are distinguished, by the greatest index, or sum of the indices of the co-ordinates, in any term of the equation.*

Thus the equation $ax = y$ belongs to a line of the *first* order, because the index of each of the co-ordinates is 1. But this order includes no *curves*. For, by art. 541, the *locus* of every such equation is a straight line.

The equation $cx^2 - axy = y^2$, belongs to the *second* order of lines, or the first kind of curves, because the greatest index is 2. The equation $ay + xy = bx$ also belongs to the second order. For, although there is here no index greater than 1, yet the *sum* of the indices of x and y, in the second term, is 2.

The equation $y^3 - 3axy = bx^2$ belongs to the *third* order of lines, or the second kind of curves, because the greatest index of y is 3.

544. In curves of the higher orders, the ordinate belonging to any given abscissa may have *different values*, and may therefore meet the curve in several points. For the length of the ordinate is determined by the *equation* of the curve, and if the equation is above the first degree, it may have two or more *roots*, (Art. 498.) and may, therefore, give different values to the ordinate.

An equation of the *first* degree has but *one* root; and a line of the first order, can be intersected by an ordinate, in one point only. Thus the equation of the line AH (Fig. 17.) is $ax = y$, in which it is evident y has but one value, while x remains the same. If the abscissa x be taken equal to AB, the ordinate y will be BD, which can meet the line AH in D only.

But the equation of the parabola, $y^2 = ax$, (Art. 530.) has *two* roots. For, by extracting both sides, $y = \pm\sqrt{ax}$. (Art. 297.) It is true that, in this case, the two values of y are *equal*. But one is *positive*, and the other *negative*. This shows that the ordinate may extend *both ways* from the end of the abscissa, and may meet the opposite branches of the curve. Thus the ordinate of the abscissa AB (Fig. 19.) may be either BD *above* the abscissa, or Bd *below* it.

A *cubic* equation has *three* roots; and an ordinate of the curve belonging to this equation, may have three different values, and may meet the curve in three different points. Thus the ordinate of the abscissa AB (Fig. 26.) may be BD, or BD', or Bd.

545. When the curve meets the axis on which the abscissas are measured, the ordinate, after becoming less and less, is reduced to nothing. (Art 533.) But, in some cases, a curve may continually approach a line, without ever meeting it. Let the distances AB, BB', $B'B'$, &c. on the line AF, (Fig. 27.) be *equal;* and let the curve $DD'D''$, &c. be of such a nature, that, of the several ordinates at the points

B, B', B'', &c. each succeeding one shall be *half* the preceding, that is, $B'D'$ half BD, $B''D''$ half $B'D'$, &c. It is evident, that, however far the straight line be carried, the curve will be coming nearer and nearer to it, and yet will never quite reach it. *A line which thus continually approaches a curve, without ever meeting it, is called an* ASSYMPTOTE *of the curve.* The axis AF is here the assymptote of the curve $DD'D''$, &c. As the abscissa increases, the ordinate diminishes, so that, when the abscissa is mathematically infinite, (Art. 447.) the ordinate becomes an infinitesimal, and may be expressed by 0. (Art. 455.)[*]

[*] See Note T.

NOTES.

NOTE A. Page 40.

IT is common to define multiplication, by saying that 'it is finding a product which has the same ratio to the multiplicand, that the multiplier has to a unit.' This is strictly and universally true. But the objection to it, *as a definition*, is, that the idea of ratio, as the term is understood in arithmetic and algebra, seems to imply a *previous* knowledge of multiplication, as well as of division. In this work, at least, geometrical ratio is made to depend on division, and division, on multiplication. Ratio, therefore, could not be properly introduced into the definition of multiplication.

It is thought, by some, to be absurd to speak of a *unit* as consisting of *parts*. But, whatever may be true with respect to number *in the abstract*, there is certainly no absurdity in considering an integer, of one denomination, as made up of parts of a different denomination. *One* rod may contain several feet; *one* foot, several inches, &c. And in multiplication, we may be required to repeat the whole, or a part of the multiplicand, as many times, as there are inches in a foot, or part of a foot.

NOTE B. p. 97.

As the *direct* powers of an integral quantity have *positive* indices, while the *reciprocal* powers have *negative* indices; it is common to call the former *positive powers*, and the latter *negative powers*. But this language is ambiguous, and may lead to mistake. For the same terms are applied to powers with positive and negative signs *prefixed*. Thus $+8a^4$ is called a positive power; while $-8a^4$ is called a negative one. It may occasion perplexity, to speak of the latter as being both positive and negative at the same time; positive, because it has a positive *index*, and negative, because

M m

it has a negative co-efficient. This ambiguity may be avoided, by using the terms direct and reciprocal; meaning, by
the former, powers with positive exponents, and, by the latter, powers with negative exponents.

Note C. p. 151.

Every affected quadratic equation may be reduced to one
of the three following forms.

$$\left.\begin{array}{ll} 1. & x^2 + ax = b \\ 2. & x^2 - ax = b \\ 3. & x^2 - ax = -b \end{array}\right\}$$

These, when they are resolved become

$$\left.\begin{array}{ll} 1. & x = -\tfrac{1}{2}a \pm \sqrt{\tfrac{1}{4}a^2 + b} \\ 2. & x = \tfrac{1}{2}a \pm \sqrt{\tfrac{1}{4}a^2 + b} \\ 3. & x = \tfrac{1}{2}a \pm \sqrt{\tfrac{1}{4}a^2 - b} \end{array}\right\}$$

In the two first of these forms, the roots are never imaginary. For the terms under the radical sign are both positive. But, in the third form, whenever b is greater than
$\tfrac{1}{4}a^2$, the expression $\tfrac{1}{4}a^2 - b$ is negative, and therefore its
root is impossible.

Note D. p. 178.

This definition of compound ratio is more comprehensive than the one which is given in Euclid. That is included
in this, but is limited to a particular case, which is stated in
art. 353. It may answer the purposes of geometry, but is
not sufficiently general for algebra.

Note E. p. 180.

It is not denied, that very respectable writers use these
terms indiscriminately. But it appears to be without any
necessity. The ratio of 6 to 2 is 3. There is certainly a
difference between *twice* this ratio, and the *square* of it, that
is, between twice three, and the square of three. All are
agreed to call the latter a *duplicate* ratio. What occasion is,

there, then, to apply to it the term *double* also? This is wanted, to distinguish the other ratio. And if it is confined to that, it is used according to the common acceptation of the word, in familiar language.

Note F. p. 185.

The definition here given is meant to be applicable to quantities of every description. The subject of proportion, as it is treated of in Euclid, is embarrassed by the means which are taken to provide for the case of *incommensurable* quantities. But this difficulty is avoided by the algebraic notation, which may represent the ratio even of incommensurables.

Thus the ratio of 1 to $\sqrt{2}$ is $\dfrac{1}{\sqrt{2}}$.

It is impossible indeed, to express, in rational numbers, the square root of 2, or the ratio which it bears to 1. But this is not necessary, for the purpose of showing its equality with another ratio.

The product $4 \times 2 = 8$.

And, as equal quantities have equal roots,

$2 \times \sqrt{2} = \sqrt{8}$, therefore, $2 : \sqrt{8} :: 1 : \sqrt{2}$.

Here the ratio of 2 to $\sqrt{8}$, is proved to be the same, as that of 1 to $\sqrt{2}$; although we are unable to find the exact value either of $\sqrt{8}$ or $\sqrt{2}$.

It is impossible to determine, with perfect accuracy, the ratio which the side of a square has to its diagonal. Yet it is easy to prove, that the side of one square has the *same* ratio to its diagonal, which the side of any other square has to its diagonal. When incommensurable quantities are once reduced to a proportion, they are subject to the same laws as other proportionals. Throughout the section on proportion, the demonstrations do not imply that we know the *value* of the terms, or their ratios; but only that one of the ratios is *equal* to the other.

$$x^2 = \dots - \alpha x$$

Note G. p. 190.

The inversion of the means can be made, with strict propriety, in those cases only in which all the terms are quantities of the same kind. For, if the two last be different from the two first, the antecedent of each couplet, after the inversion, will be different from the consequent, and therefore, there can be no ratio between them. (Art. 355.)

This distinction, however, is of little importance in practice. For, when the several quantities are expressed in *numbers*, there will always be a ratio between the numbers. And when two of them are to be multiplied together, it is immaterial which is the multiplier, and which the multiplicand. Thus, in the Rule of Three in arithmetic, a change in the order of the two middle terms will make no difference in the result.

Note H. p. 197.

The terms *composition* and *division* are derived from geometry, and are introduced here, because they are generally used by writers on proportion. But they are calculated rather to perplex, than to assist, the learner. The objection to the word *composition* is, that its meaning is liable to be mistaken for the composition or compounding of *ratios*. (Art. 390.) The two cases are entirely different, and ought to be carefully distinguished. In one, the terms are *added*, in the other, they are *multiplied* together. The word compound has a similar ambiguity in other parts of the mathematics. The expression $a+b$, in which a is *added* to b, is called a compound quantity. The fraction $\frac{1}{2}$ of $\frac{2}{3}$, or $\frac{1}{2} \times \frac{2}{3}$, in which $\frac{1}{2}$ is *multiplied* into $\frac{2}{3}$, is called a compound fraction.

The term *division*, as it is used here, is also exceptionable. The alteration to which it is applied, is effected by *subtraction*, and has nothing of the nature of what is called division in arithmetic and algebra. But there is another case, (Art. 392.) totally distinct from this, in which the change in the terms of the proportion is actually produced by division.

Note I. p. 203.

The principles stated in this section, are not only expressed in different language, from the corresponding propositions in Euclid, but are, in several instances, more general. Thus the first proposition in the fifth book of the Elements, is confined to *equimultiples*. But the article referred to, as containing this proposition, is applicable to all cases of equal *ratios*, whether the antecedents are multiples of the consequents or not.

Note K. p. 217.

The solution of one of the cases is omitted in the text, because it is performed by *logarithms*, with which the learner is supposed not to be acquainted, in this part of the course. When the first term, the last term, and the ratio, are given, the *number* of terms may be found by the formula

$$n = \frac{\log. \frac{rz}{a}}{\log. r}.$$

Note L. p. 221.

When it is said that a mathematical quantity may be supposed to be increased beyond any determinate limits, it is not intended that a quantity can be specified so great, that no limits greater than this can be assigned. The quantity and the limits may be *alternately* extended one beyond the other. If a line be conceived to reach to the most distant point in the visible heavens, a limit may be mentioned beyond this. The line may then be supposed to be extended farther than this limit. Another point may be specified still farther on, and yet the line may be conceived to be carried beyond it.

Note M. p. 223.

The apparent *contradictions* respecting infinity, are owing to the ambiguity of the term. It is often thought that the proposition, that quantity is infinitely divisible, involves an absurdity. If it can be proved that a line an *inch* long can be divided into an infinite number of parts, it can, by the same mode of reasoning, be proved, that a line *two inches* long may be first divided in the middle, and then each of the sections be divided into an infinite number of parts. In this way, we shall obtain one infinite *twice as great* as another.

If by infinity, here, is meant that which is beyond any assignable limits, one of these infinites may be supposed greater than the other, without any absurdity. But if it be meant that the number of divisions is so great that it can not be increased, we do not prove this, concerning either of the lines. We make out, therefore, no contradiction. The apparent absurdity arises from shifting the meaning of the terms. We demonstrate that a quantity is, in one sense, infinite; and then infer that it is infinite, in a sense widely different.

Note N. p. 227.

Strictly speaking, the inquiry to be made is, how often the *whole* divisor is contained in as many terms of the dividend. But it is easier to divide by a *part* only of the divisor; and this will lead to no errour in the result, as the whole divisor is multiplied, in obtaining the several subtrahends.

Note O. p. 235.

The demonstration of this proposition, particularly in its application to fractional indices, could not be introduced, with advantage, in this part of the course. It does not appear that Newton himself demonstrated his theorem, except by induction. And though various demonstrations have since been given; yet they are generally founded upon principles and methods of investigation not contained in this introduction, such as the laws of combinations, fluxions, and figurate numbers.

Those who wish to examine the inquiries on this subject, may consult Simpson's Algebra, Section 15, Euler's Algebra, Section II. Chap. 11, Vince's Fluxions, Art. 99, Lacroix's Algebra, Art. 138, &c. Do. Comp. Art. 71, Rees' Cyclopedia, Manning's Algebra, and the London Phil. Trans. Vol. xxxv. p. 298.

Note P. p. 253.

The very limited extent of this work would admit of nothing more, than a *bare specimen* of the Summation of Series. For information on this subject, the learner is referred to Emerson's Method of Increments, Sterling's Summation of Series, Waring's Fluxions, Maclaurin's Fluxions, Art. 828, &c. Wood's Algebra, Art. 410, Lacroix's Comp. Alg. Art. 81, &c. Euler's Anal. Infin. C. XIII. Simpson's Essays and Dissertations, De Moivre's Miss. Analyt. p. 72, and the London Philosophical Transactions.

Note Q. p. 261.

To those who have made any considerable progress in the mathematics, this section will doubtless appear very defective. But it was impossible to do juctice to the subject, without occupying more room, than could be allotted to it here. In going through an elementary course of mathematics and natural philosophy, the student will rarely have occasion to solve an equation above the second degree.

Those who wish to examine particularly the different methods of solution, will find them, in Newton's Universal Arithmetic, Maclaurin's Alg. Part II, Euler's Alg. Part 1. Sec. 4, Waring's Algebra, Do. Medit. Algeb. Wallis' Algebra, Simpson's Alg. Sec. 12, Fenn's Alg. Ch. 3 and 4. Saunderson's Alg. Book x, Simpson's Essays and Dissertations, Journal De Physique, Mar. 1807, and Philosophical Transactions.

Note R. p. 267.

It will be thought, perhaps, that it was unnecessary to be so particular, in obtaining the expression for the area of a parallelogram, for the use of those who read Playfair's edition of Euclid, in which "*AD.DC* is put for the rectangle contained by *AD* and *DC*." It is to be observed, however, that he introduces this, merely as an article of *notation*. (Book II. Def. 1.) And though a point interposed between the letters, is, in algebra, a sign of multiplication; yet he does not here undertake to show how the sides of a parallelogram may be multiplied together. In the first book of the *Supplement*, he has indeed demonstrated, that "equiangular parallelograms are to one another, as the products of the numbers proportional to their sides." But he has not given to the expressions, the forms most convenient for the succeeding parts of this work. In making the transition from pure geometry to algebraic solutions and demonstrations, it is important to have it clearly seen, that the geometrical principles are not altered; but are only expressed in a different language.

Note S. p. 275.

This section comprises very little of what is commonly understood by the application of algebra to geometry. The principal object has been, to prepare the way for the other parts of the course, by stating the grounds of the algebraic notation of geometrical quantities, and rendering it familiar by a few examples.

On the construction and solution of problems, see Newton's Arithmetic, Simpson's Alg. Sec. 18 and appendix, Lacroix's App. Alg. Geom. Saunderson's Alg. Book XIII, Analyt. Instit. of Maria Agnesi, Book 1, Sec. 2, and Emerson's Alg. Book II. Sec. 6.

Note T. p. 288.

On the equations of curves, the geometrical construction of equations, the finding of *loci*, &c. see Maclaurin's Alg. Part III, and appendix, Newton's Arith. Emerson's Alg. Book II. Sec. 9, Do. Prop. of Curves, Euler's Anal. Infin. Waring's Prop. Alg. and Mansfield's Essays.

Among the subjects which, for want of room, are entirely omitted in this introduction, one of the most interesting is the *indeterminate analysis*. No part of algebra, perhaps, is better calculated to exercise the powers of *invention*. But other branches of the mathematics are so little dependent on this, that it is not absolutely necessary to give it a place in an elementary course.

See, on this subject, Euler's Algebra, Vol. II, with Lagrange's additions, Saunderson's Alg. Book VI, and the Edinburgh Phil. Transactions, Vol. II.

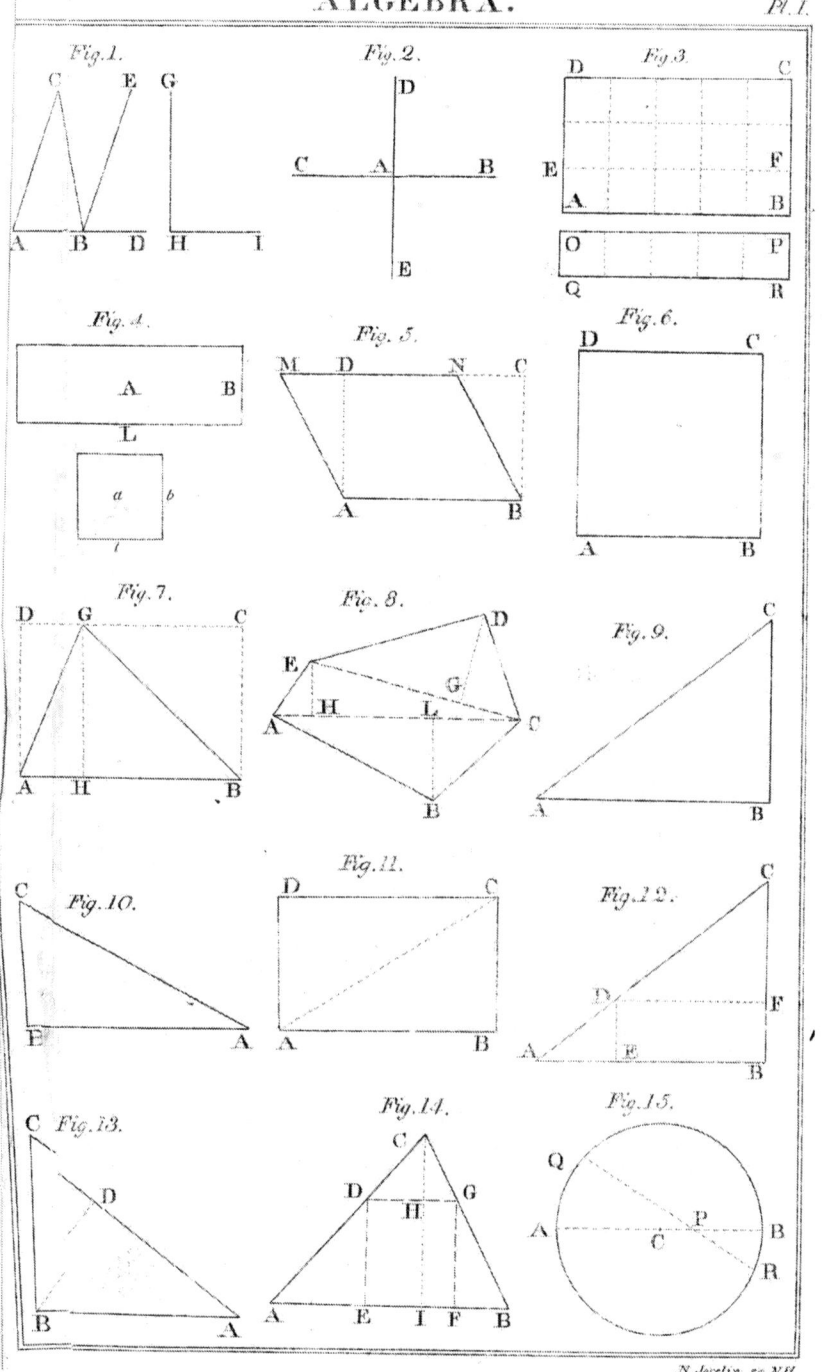

succeeding parts of this work. In making the transition from pure geometry to algebraic solutions and demonstrations, it is important to have it clearly seen, that the geometrical principles are not altered; but are only expressed in a different language.

Note S. p. 275.

This section comprises very little of what is commonly understood by the application of algebra to geometry. The principal object has been, to prepare the way for the other parts of the course, by stating the grounds of the algebraic notation of geometrical quantities, and rendering it familiar by a few examples.

On the construction and solution of problems, see Newton's Arithmetic, Simpson's Alg. Sec. 18 and appendix, Lacroix's App. Alg. Geom. Saunderson's Alg. Book xiii, Analyt. Instit. of Maria Agnesi, Book 1, Sec. 2, and Emerson's Alg. Book ii. Sec. 6.

Note T. p. 288.

On the equations of curves, the geometrical construction of equations, the finding of *loci*, &c. see Maclaurin's Alg. Part iii, and appendix, Newton's Arith. Emerson's Alg. Book ii. Sec. 9, Do. Prop. of Curves, Euler's Anal. Infin. Waring's Prop. Alg. and Mansfield's Essays.

Among the subjects which, for want of room, are entirely omitted in this introduction, one of the most interesting is the *indeterminate analysis*. No part of algebra, perhaps, is better calculated to exercise the powers of *invention*. But other branches of the mathematics are so little dependent on this, that it is not absolutely necessary to give it a place in an elementary course.

See, on this subject, Euler's Algebra, Vol. ii, with Lagrange's additions, Saunderson's Alg. Book vi, and the Edinburgh Phil. Transactions, Vol. ii.

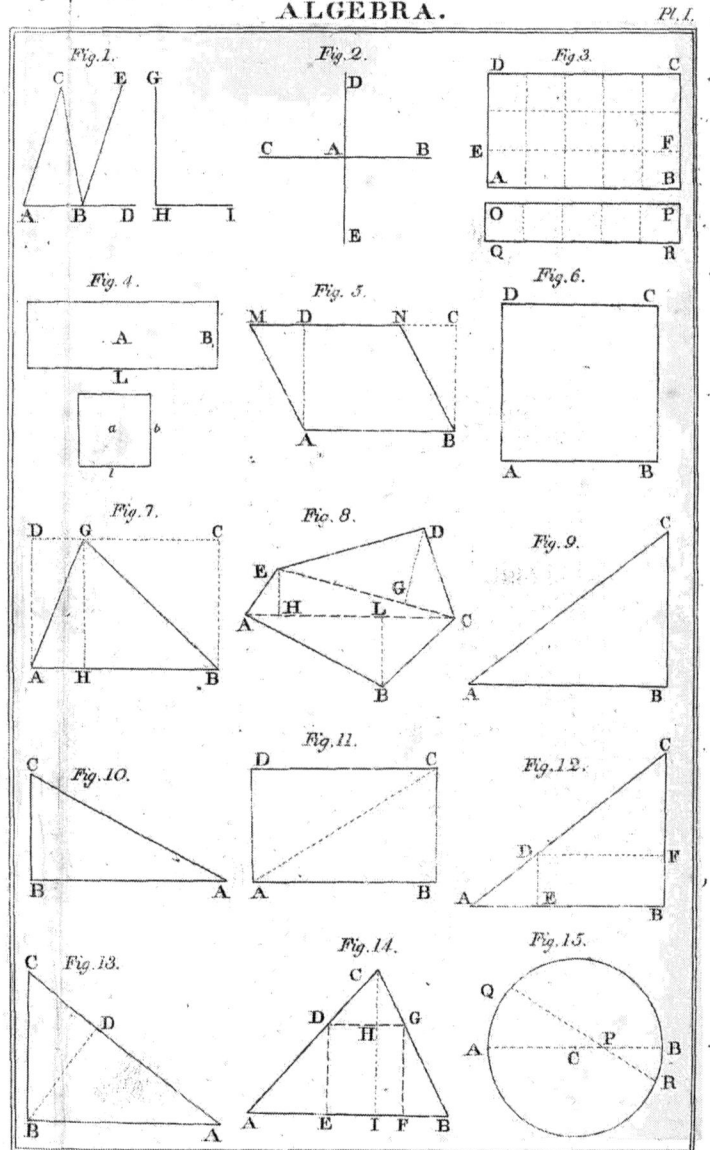

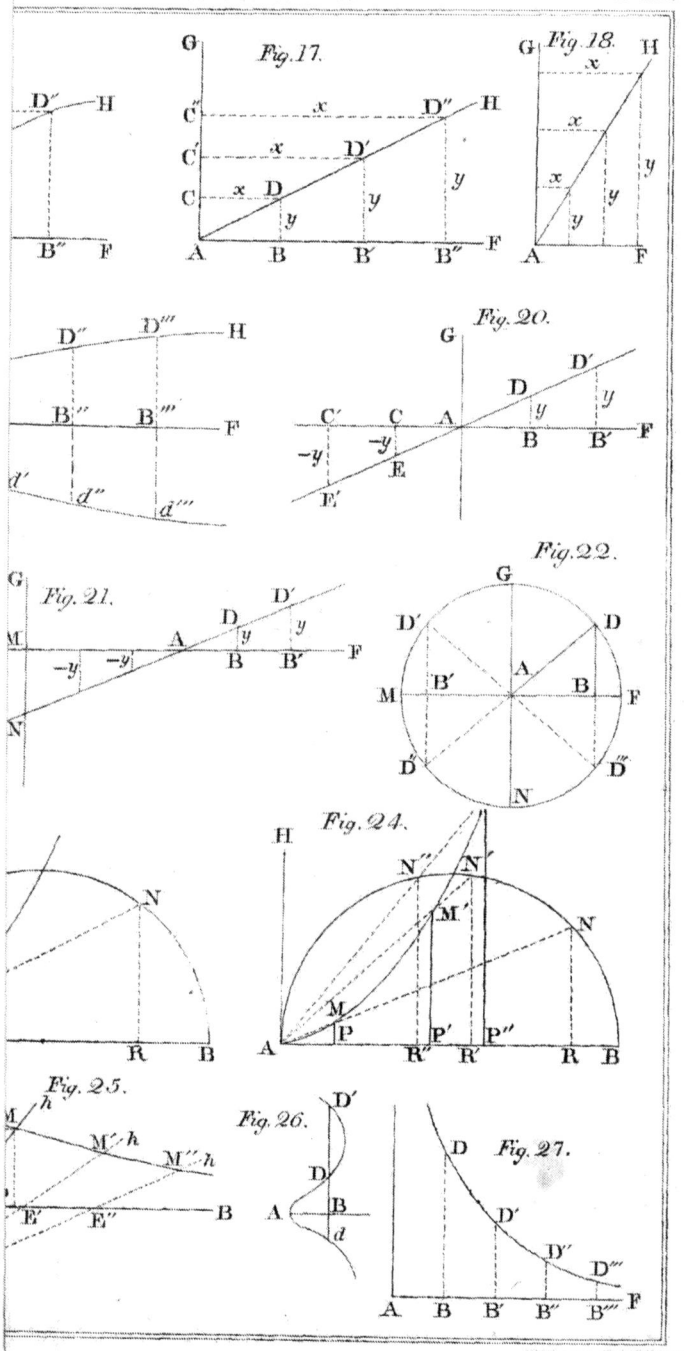

Fig. 17.

Fig. 18.

Fig. 20.

Fig. 21.

Fig. 22.

Fig. 24.

Fig. 25.

Fig. 26.

Fig. 27.

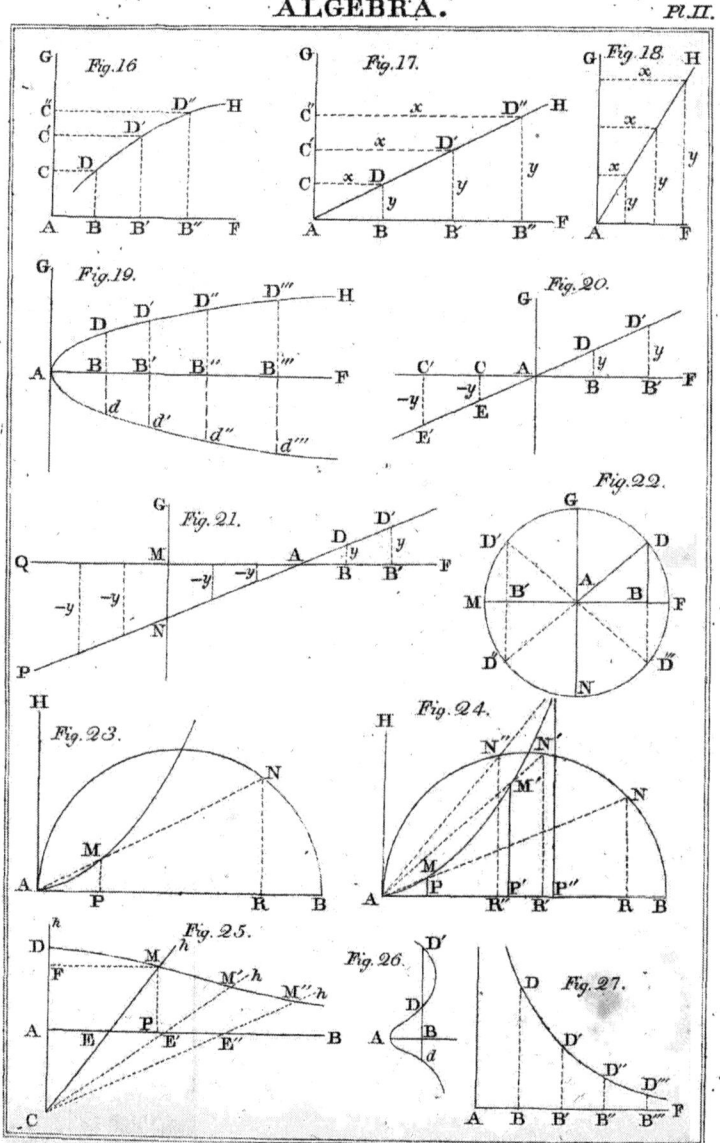

Lightning Source UK Ltd.
Milton Keynes UK
UKHW021421060122
396675UK00003B/122

9 781015 380998